FROM HERE TO
INFINITY

∞

Also by Martin Rees

Before the Beginning

Our Final Hour

Our Cosmic Habitat

Just Six Numbers

FROM HERE TO INFINITY

A VISION FOR
THE FUTURE OF SCIENCE

Martin Rees

W. W. NORTON & COMPANY

New York • London

For information about special discounts for bulk purchases, please contact
W. W. Norton Special Sales at specialsales@wwnorton.com or 800-233-4830

Manufacturing by Courier Westford
Book design by Chris Welch
Production manager: Julia Druskin

Library of Congress Cataloging-in-Publication Data

Rees, Martin J., 1942–
From here to infinity : a vision for the future of science /
Martin Rees. — 1st American ed.
p. cm.
ISBN 978-0-393-06307-3 (hardcover)
1. Science—Philosophy. 2. Science and civilization. I. Title.
Q171.R34 2012
500—dc23
2012003421

W. W. Norton & Company, Inc.
500 Fifth Avenue, New York, N.Y. 10110
www.wwnorton.com

W. W. Norton & Company Ltd.
Castle House, 75/76 Wells Street, London W1T 3QT

1234567890

CONTENTS

INTRODUCTION

Science impinges on us more than ever before: its findings deepen our understanding of the world and what lies beyond; its applications transform our lives. These transformations are mainly beneficial, but often open up threats and ethical challenges. Indeed they could threaten us all. This century is the first in the Earth's history (spanning 45 million centuries) when one species, ours, has the power to determine the fate of the entire biosphere. These were my themes in the BBC Reith Lectures, which I was privileged to give in 2010 and on which this short book is based.

Questions about human origins, life in space, our long-range destiny, and the laws of nature fascinate a wide public. I'd enjoy my own research work less if I couldn't share its outcome with nonspecialists—not only what we've learned, but our attempts to tackle the big questions that remain unanswered (and some of which may forever be beyond human grasp). The words of the Anglo-American philosopher A. N. Whitehead are as true today as ever: "Philosophy begins in wonder, and at the end, when philosophic thought has done its best, the wonder remains."

There is another reason, quite apart from their intrinsic fascination, why the key concepts of science need to percolate widely: without a feel for science (and how to assess risks and uncertainty) we cannot as citizens participate in the increasing range of key political issues—energy, health, environment, and so forth—that have a scientific dimension.

Some might think that intellectual immersion in vast expanses of space and time would render cosmologists serene and uncaring about what happens next year, next week, or tomorrow. But for me the opposite is the case. My concerns deepen with the awareness that, even in a perspective extending billions of years into the future, as well as into the past, this century may be a defining moment: there's a genuine risk that our actions could jeopardize not only the immediate future but also life's immense potential. That's why I see no incongruity in combining, in the chapters that follow, some speculations about scientific frontiers with discussions of practical issues of policy and ethics.

The Reith Lectures were established in honor of John Reith, a formidable and austere Scotsman who was the first head of the BBC. Right from the start, he envisioned the role of broadcasting as being "to inform, educate and entertain." Even today, with the media world hugely diversified and utterly transformed, the BBC remains perhaps the preeminent public service broadcasting organization, with a worldwide reach.

The philosopher Bertrand Russell gave the first Reith lecture in 1948. Since then a succession of figures from culture, science, and public affairs have presented lectures, including several distinguished Americans—from J. Robert Oppenheimer (1953) and George F. Kennan (1957) to the Harvard philosopher Michael Sandel (2009).

The themes of my own lectures are, I believe, as relevant to US readers as to those in Britain. Science itself is, after all, the one truly global culture, transcending all boundaries of nationality and faith. Moreover, threats like climate change and pandemics don't just concern a single nation; all major geopolitical groupings—not only the United States but also the European Union and the emerging Asian superpowers—need to act in concert. Indeed, even in some areas of "pure" science, no single country can go it alone: the biggest accelerators and space projects have to be planned on a world scale.

The United States is the number-one nation in the quality and volume of its science. And there is a flood of excellent material presenting and interpreting science to nonspecialists. Indeed I, myself, was inspired as a young scientist by the example of Carl Sagan, who tried to break down boundaries through his talks, articles, and media appearances, especially his TV series *Cosmos*. He enlightened millions of viewers from around the world and sensitized them to the hopes and fears science raises.

A disclosure: had my original audience been primarily American, I would have said more about the prospects and rationale for space exploration. The scale of the programs of NASA (and of the US military) were ramped up by superpower rivalry during the Cold War. Everyone remembers Neil Armstrong's "one small step" on the Moon—an inspirational antidote to the regular news of carnage in Vietnam during the same years—but the momentum that triggered this achievement soon dissipated. European activities in space are more modest, despite the overall economic and intellectual parity between Europe and the United States. I don't think Europe should initiate a manned program of its own: sending people back to the Moon, or beyond, can't be justified except as a high-risk adventure or spectator sport. Instead, Europe's emphasis should be on an unmanned program, spearheaded by robotic probes and fabricators.

We Europeans view with admiration the phalanxes of well-informed and politically engaged scientists in the United States—especially in areas of defense and arms control, where most of the leading independent voices have come from America's academic community. In fact, America's world-leading scientific expertise provides unrivaled professionalism in addressing the problems of health, energy, population, and environmental change. And there is no lack of policy prescriptions jostling for the attention of US politicians. President Obama appointed a "dream team" of internationally respected scientists to key posts in his administration. But there is one impediment that engenders equal frustration on both sides of the Atlantic—indeed in all democracies: a tendency for long-term strategies, however important, to be trumped by more immediate issues that can be resolved within an electoral cycle.

Both the United States and Britain worry about the quality of their schools—and the unequal access that young people have to excellent teaching, especially in science. We confront a growing challenge from Asia, where pupils' attainment levels are higher. These issues have been addressed in the United States by the National Academy of Sciences, and in Britain by the Royal Society (of which I was until recently the president).

Despite all we share, though, there are differences in priorities and attitudes between the United States and Britain (and Western Europe). On some issues (gun laws and capital punishment, for instance) the gulf between median opinion on the two sides of the Atlantic is very wide. And it is paradoxical that in the United States, the country that is preeminent in its volume of scientific excellence, antiscientific sentiment seems most widespread and politically influential. Controversies over stem cell and embryo research and the teaching of creationism have been widespread and high-profile in the United States. In Britain, these conflicts have been less acute—a consequence, perhaps, of its being a more secular society, where mainstream religious leaders are supportive of science and fundamentalism is weaker. Despite the intemperate rhetoric of a combative atheistic fringe, the majority of scientists in Britain, whether or not they themselves adhere to any faith, are content to maintain peaceful coexistence with organized religion. And they would acknowledge that decisions on how science should be applied—and choices between what can be done and what should be done—entail ethical judgments that science alone can't provide. My main aim in the Reith Lectures was to offer a personal perspective on these challenges.

A WORD ABOUT the format of this book. My four lectures were on a common theme, but each had to be self-contained and to offer sufficient variety to sustain the interest of a disparate audience. Moreover, they were recorded in front of a live audience. In converting them into a book I have expanded and updated some of the material—putting in what I would have said if each lecture had been longer, and also taking account of comments made by the live audiences and by listeners who sent letters or e-mails. However I have tried to keep the flavor and style of the lectures; in particular, each chapter can be read independently, even though this means occasional repetition. Overall, the lectures ranged from timely issues in science policy to the frontiers and limits of our knowledge, and the (perhaps infinite) universe itself. No topics could be explored in depth: the treatment of each was inevitably sketchy and telegraphic. The most I can aspire to is that readers of wide-ranging tastes find the outcome closer to a smorgasbord than a dog's breakfast.

The lectures were recorded in different locations, and I tried to slant the theme of each toward the likely interests of those in the live audience. The opening lecture was in the BBC's own Lecture Theater at Broadcasting House, and in those surroundings it seemed appropriate to focus on how scientists interact with politicians and the public. And the final lecture, given at the Open University (a world leader in distance learning and a public nonprofit venture), addressed the impact on science and education of globalization and modern communications, and what it's like to be a scientist. In Cardiff, where the audience was specially diverse, I spoke about something that concerns us all: what the world might be like in 2050. The lecture held at the Royal Society was the only one that attracted a predominantly scientific audience. Here I was somewhat more self-indulgent in discussing future developments in science itself. It might have been more appropriate for these "cosmic" and speculative comments to round off the series, but for scheduling and logistic reasons this lecture was the third rather than the fourth, and I have preserved this ordering in the chapters of the present book.

Reith Lectures have on several occasions been given by scientists or engineers; the most formidable and eloquent among them was the late Peter Medawar, who spoke in 1959 on "The Future of Man." He speculated on future trends in biology and genetics, his tone being optimistic despite his awareness of the downsides. And his concluding sentence is one that I would echo, fifty years later and with greater urgency:

> The bells which toll for mankind are—most of them, anyway—like the bells on Alpine cattle; they are attached to our own necks, and it must be our fault if they do not make a cheerful and harmonious sound.

1

THE SCIENTIFIC
CITIZEN

et us start with a flashback to the 1660s—to the earliest days of one of the world's first scientific academies, the Royal Society of London. Christopher Wren, Robert Hooke, Samuel Pepys, and other "ingenious and curious gentlemen" (as they described themselves) met regularly. Their motto was to accept nothing on authority. They did experiments; they peered through newly invented telescopes and microscopes; they dissected weird animals. But, as well as indulging their curiosity, they were immersed in the practical agenda of their era: improving navigation, exploring the New World, and rebuilding London after the Great Fire.

Today, science has transformed our lives. Our horizons have hugely expanded; no new continents remain to be discovered. Our Earth no longer offers an open frontier, but seems constricted and crowded—a "pale blue dot" in the immense cosmos. Despite all that has changed, the values and attitudes of the Royal Society's founders are enduring ones. Today's scientists are specialists, not polymaths. But, like their forebears, they probe nature and nature's laws by observation and experiment; and they should also engage broadly with society and with public affairs.

Indeed, their engagement is needed more than ever before. But science isn't just for scientists. We should all have a voice in ensuring that it's applied ethically, and to the benefit of both the developing and the developed world. We must confront widely held anxieties that genetics, brain science, and artificial intelligence may "run away" too fast. As citizens, we all need a feel for how much confidence can be placed in science's claims.

Science as culture

Quite apart from its pervasive impact on our lives, science is part of our culture. In 2009 Charles Darwin's anniversary was celebrated—especially, of course, in his home country, but around the world. Darwin's impact on nineteenth-century thought was profound, but it resonates even more today. His concept of natural selection was described by Dan Dennett, with only slight hyperbole, as "the best idea anyone ever had." His insights are pivotal to our understanding of all life on Earth, and the vulnerability of our environment to human actions. Other sciences have disclosed the nature of atoms, DNA, and the stars, and have enlarged our cosmic horizons. It's a *cultural* deprivation not to appreciate the panorama offered by modern cosmology and Darwinian evolution—the chain of emergent complexity leading from some still-mysterious beginning to atoms, stars, and planets—and how, on our planet, life emerged and evolved into a biosphere containing creatures with brains able to ponder the wonder of it all. This common understanding should transcend all national differences—and all faiths too.

Science is indeed a global culture. Its universality is specially compelling in my own subject of astronomy. The dark night sky is an inheritance we've shared with all humanity, throughout history. All have gazed up in wonder at the same vault of heaven, but interpreted it in diverse ways. There is a natural fascination with the big questions: Was there a beginning? How did life emerge? Is there life in space? And so forth.

The simplest building blocks of our world—atoms—behave in ways physicists can understand and calculate. And the laws and forces governing them are universal: atoms behave the same way everywhere on Earth—indeed they are the same even in the remotest stars. We know these basics well enough to enable engineers to design all the mechanical artifacts of our modern world, from radios to rockets. Our environment is far too complicated for every detail to be explained but our perspective has been transformed by great insights—great unifying ideas. The concept of continental drift, for instance, helps us to fit together a whole raft of geological and ecological patterns across the globe. And Darwin's great insight revealed the overarching unity of the entire web of life on our planet.

There are patterns in nature. There are even patterns in how we humans behave—in how cities grow, how epidemics spread, and how technologies like silicon chips develop. The more we understand the world, the less bewildering it becomes, and the more we're able to change it.

These "laws" or patterns are the great triumphs of science. To discover them required dedicated talent—even genius. But to grasp their essence isn't so difficult: most of us appreciate music even if we can't compose or perform it. Likewise, the key ideas of science can be accessed and enjoyed by almost everyone; the technicalities may be daunting, but these are less important for most of us and can be left to the specialists.

Politics and scientific uncertainty

The twenty-first century is a crucial one. The Earth has existed for 45 million centuries but this is the first period in which one species, ours, can determine, for good or ill, the future of the entire biosphere. Over most of history, the threats have come from nature—disease, earthquakes, floods, and so on. But now they come from us. We've entered what some have called the "Anthropocene" era.

But despite the concerns there are powerful grounds for optimism. For most people in most nations, there's never been a better time to be alive. The innovations driving economic advance—information technology, biotechnology, and nanotechnology—can boost the developing as well as the developed world. Creativity in science and the arts is nourished by a wider range of influences—and is accessible to hugely more people worldwide—than in the past. We're becoming embedded in a cyberspace that can link anyone, anywhere, to all the world's information and culture, and to most other people on the planet. Twenty-first-century technologies can offer everyone a lifestyle that requires little compromise on what Americans and Europeans aspire to today, while being environmentally benign, involving lower demands on energy or resources.

Some changes happen with staggering speed. Everyday life has been transformed in less than two decades by mobile phones and the Internet. Computers double their power every two years. Spin-offs from developments in genetics could soon be as pervasive as those we've already seen from the microchip. These rapid advances, and others across the whole of science, raise profound questions. Who should access the readout of our personal genetic code? How will our lengthening life spans affect society? Should we build nuclear power stations, or wind farms, if we want to keep the lights on? Should we use more insecticides, or plant genetically modified (GM) crops? Should the law allow "designer babies"? How much should computer technology be permitted to invade our privacy?

Such questions don't feature much in national election campaigns in any country. That's partly because they transcend party politics, but it's also because they're long-term, and tend to be trumped by more urgent items on political agendas. But often science has an *urgent* impact on our lives. Governments and businesses, as well as individuals, then need specialist advice—advice that fairly presents the level of confidence and the degree of uncertainty.

Issues come up unexpectedly. For instance, in April 2010, the volcanic eruptions in Iceland that disrupted air travel in Northern Europe raised urgent questions about vulcanology, about wind patterns, and about how volcanic dust affects jet engines. In that instance, the knowledge was, basically, there; what was lacking was coordination, and an agreement on the acceptable level of risk. After this episode, a code of practice was agreed that should ensure that future events of this kind are handled more smoothly. Sometimes, though, the key science isn't known. An interesting example was the outbreak of BSE or "mad cow disease" in Britain in the 1980s. At first, experts conjectured that this disease posed no threat to humans because it resembled scrapie in sheep, which had been endemic for 200 years without crossing the species barrier. That was a reasonable conjecture, and comforting to politicians and the public, but it proved wrong. The pendulum then swung the other way. Banning "beef on the bone," for instance, was, in retrospect, an overreaction, but at the time seemed a prudent precaution against a potential tragedy that could have been far more widespread than it actually turned out to be.

Likewise, governments were prudent to stock up vaccines against swine flu—even though, fortunately, the epidemics have so far proved milder than feared. Indeed, if we apply to pandemics the same prudent analysis whereby we calculate an insurance premium—multiplying probability by consequences—we'd surely conclude that measures to alleviate this kind of extreme event should actually be scaled up. (And these measures need international cooperation. Whether or not an epidemic gets a global grip may hinge, for instance, on how quickly a Vietnamese poultry farmer can report any strange sickness.)

Incidentally, there's a mismatch between public perception of very different risks and their actual seriousness. We fret unduly about carcinogens in food and low-level radiation. But we are in denial about low-probability, high-consequence events that should concern us more. The recent financial crash was one such event; but others that haven't yet happened—lethal pandemics are one example—should loom higher on the agenda.

The varied topics mentioned above show how pervasive science is, in our lives and in public policy. President Obama certainly recognized this when he filled some key posts in his administration with a dream team of top-rate scientists. He opined that their advice should be heeded "even when it is inconvenient—indeed especially when it is inconvenient." In Britain we have a chief science advisor, and separate independent advisors in most government departments.

Winston Churchill valued advice from scientists (which was plainly crucial in World War II), but kept them in their place; he insisted that they should be "on tap, not on top." It is indeed the elected politicians who should make decisions. But the role of scientific advice is not just to provide facts, still less to support policies already decided. Experts should be prepared to challenge decision-makers, and help them to navigate the uncertainties. But there's one thing that scientific advisors mustn't forget. Whether the context be nuclear weapons, nuclear power, drug classification, or health risks, political decisions are seldom *purely* scientific: they involve ethics, economics, and social policies as well. And in domains beyond their special expertise, scientists speak just as citizens, with no enhanced authority.

There's no denying where science has recently had the most contentious policy impact, and where the long-term stakes are highest: climate change. Climate science is complex, involving a network of intermeshing effects. But there is one decisive piece of evidence: the amount of carbon dioxide in the atmosphere is higher than it's been for at least half a million years, and it is inexorably rising, mainly because of the burning of fossil fuels. This measurement *isn't* controversial. And straightforward chemistry tells us that carbon dioxide is a so-called greenhouse gas; it acts like a blanket, preventing some of the heat radiated by the Earth from escaping freely into space. So the carbon dioxide buildup in the atmosphere will trigger a long-term warming, superimposed on all the other complicated effects that make climate fluctuate.

What is uncertain, however, is the *predicted rate* of warming—calculations yield a spread of projections, depending on how much the poorly understood "feedback" from water vapor and clouds enhances the blanketing. The high projections are of course more threatening than the low ones, and so much is at stake that it's crucial to develop better understanding so that predictions can be firmed up. Nonetheless, even the existing science convinces me that the threat of seriously disruptive climate change is significant enough to justify its priority on the political agenda.

This confidence may surprise anyone who has dipped into all that's been written on the subject. Any trawl of the Internet reveals diverse and contradictory claims. How do you make up your mind? The following analogy suggests an answer.

Suppose you seek medical guidance. Googling any ailment reveals a bewildering range of purported remedies.

But, if your own health were at stake, you wouldn't attach equal weight to everything in the blogosphere: you'd entrust yourself to someone with manifest medical credentials and a successful record of diagnosis. Likewise, we get a clearer "steer" on climate—though not, of course, a complete consensus—by attaching more weight to those with serious credentials in the subject. But, as already noted, it's crucial to keep "clear water" between the science on the one hand and the policy response on the other. Risk assessment should be separate from risk management.

Even if there were *minimal* uncertainty about how the world's weather might change, there would still be divergent views on what governments should do about it. Climate scientists would themselves still have a range of opinions on what the best policies were: but they should express these views as citizens, and not claim any special weight for their policy judgments.

There's a balance to be struck between mitigating climate change and adapting to it. And there are other questions. How much should we sacrifice now to ensure that the world is no worse when our grandchildren grow old? How much subsidy should be transferred from the rich world, whose fossil fuel emissions have mostly caused the problem, to the developing nations? How much should we incentivise clean energy? Should we gamble that our successors may devise a technical fix that will render nugatory any actions we take now? On all these choices there's as yet minimal consensus, still less effective action. But policies, and investment priorities, are being influenced by climate change projections. So it's inevitable, and right, that climate science is under specially close scrutiny.

In the past, most people unquestioningly accepted "authorities" on any topic, but that has now changed. We can all access far more information than ever before and we want to weigh up evidence for ourselves. Such scrutiny should be welcome; just as there are instances of shoddy work, error, or even malpractice in the medical and legal professions, so there are occasionally in science.

Current practice in archiving and managing data is not uniform across all fields, nor across all countries. Nor is there a consensus on the appropriate guidelines for making such information available. Under the United Kingdom's Freedom of Information Act, anyone, whether a UK taxpayer or not, whether they have good reason or not, can impose burdensome demands on researchers by repeated requests. It is not obvious that this is right. Similar issues arise in the United States, where members of Congress have made vexacious requests for data from researchers, or for access to e-mails. On the other hand, we surely need to facilitate open debate, to ensure that scientific claims are robust and firmly grounded.

Scientists are their own severest critics. They have more incentive than anyone else to uncover errors. That's because the greatest esteem goes to those who contribute something unexpected and original, and especially to those who can overturn a consensus. That's how initially tentative ideas get firmed up—not only on climate change, but (to cite some examples from earlier years) regarding the link between smoking and lung cancer, and between HIV and AIDS. But that's also how seductive theories get destroyed by harsh facts. Science is organized skepticism.

The path toward a consensual understanding is often winding, with many blind alleys being explored before reaching it. Sometimes, a prior consensus is overturned—though Thomas Kuhn's famous book on scientific revolutions perhaps exaggerates how often this happens. The Copernican cosmology, overthrowing the concept of a geocentric cosmos, would qualify as a genuine revolution, as would quantum theory. But most advances transcend and generalize the concepts that went before, rather than contradict them. For instance, Einstein didn't overthrow Newton. His work led to a theory that had broader scope and gave deeper insights into the nature of space and gravity, but Newton's laws are still good enough to predict the trajectories of spacecraft. (There is, incidentally, one practical context where Einstein's refinements are needed: the accuracy of the Global Positioning Satellites (GPS) used in SatNav systems would be fatally degraded if proper allowance wasn't made for the slight difference between the clock rates on Earth and those in orbit that is predicted by relativity theory.)

As a student at Cambridge University in the 1960s I watched at close hand a standoff between Martin Ryle and Fred Hoyle—two outstanding scientists utterly different in their personal and professional styles—on whether the universe had emerged from a big bang or whether it had existed forever in a so-called steady state. New evidence settled this debate in Ryle's favor, an outcome to which Hoyle was never fully reconciled, though by the end of his life he was advocating a compromise "steady bang" theory that gained little traction with others. Likewise, there have been high-profile vendettas on conceptually important issues in evolutionary biology and sociobiology. There are fewer purely intellectual disputes today that become so deeply personalized. This is not due to the sweeter dispositions of a younger generation of scientists, but because, as data accumulate, there is progressively less scope for viable but strongly divergent hypotheses; and there is a growing incentive toward collaborative rather than isolated research.

When rival theories fight it out there is eventually just one winner—at most. Sometimes, one crucial piece of evidence clinches the case, as happened for the big bang cosmology and also for continental drift. In other cases, an idea gains only a gradual ascendancy as alternative views get marginalized until their leading proponents die off. Sometimes, the subject moves on, and what once seemed an epochal issue is bypassed or sidelined.

Our scientific knowledge and capabilities are actually

surprisingly patchy. Odd though it may seem, some of the best-understood phenomena are far away in the cosmos. Back in the seventeenth century, Newton could describe the "clockwork of the heavens"; eclipses could be both understood and predicted. (Indeed, even in Babylonian times, regularities and repetitions were discerned and some prediction was possible even in ignorance of the underlying mechanism.) But few other things are so predictable. For instance, it's hard to forecast, even a day before, whether those who go to view an eclipse will encounter clouds or clear skies. And our understanding of some familiar matters that interest us all—diet and child care, for instance—is still so meager that expert advice changes from year to year. Everyday phenomena, especially those involving entities as complex as human beings, can be more intractable than anything in the inanimate world.

If you ask scientists what they are working on, you will seldom get an inspirational reply like "seeking to cure cancer" or "understanding the universe." Rather, they will focus on a tiny piece of the puzzle and tackle something that seems tractable. Scientists are not thereby ducking the big problems, but judging instead that an oblique approach can often pay off best. A frontal attack on a grand challenge may, in fact, be premature. For instance, forty years ago President Richard Nixon declared a war on cancer, envisaging this as a national goal modeled on the then-recent Apollo Moon-landing program. But there was a crucial difference. The science underpinning *Apollo*—rocketry and celestial mechanics—was already understood, so that, when funds gushed at NASA, the Moon landings became reality. But in the case of cancer the scientists knew too little to be able to target their efforts effectively.

It's easy to think of other examples. Suppose that a nineteenth-century innovator had wanted to develop better machines to reproduce music. He could have made very elaborate mechanical organs or pianolas, but wouldn't have accelerated the advent of radio or the MP3 player. And a medical program to seek ways to see through flesh certainly wouldn't have stimulated the serendipitous discovery of X-rays.

The word "science" is being used here in a broad sense, by the way, to encompass technology and engineering; this is not just to save words, but because all of these disciplines are symbiotically linked. The mental process of problem solving motivates us all, whether one is an astronomer probing the remote cosmos or an engineer facing a down-to-earth design conundrum. There is at least as much challenge in the latter, a point neatly made by an old cartoon showing two beavers looking up at a hydroelectric dam. One beaver says, "I didn't actually build it, but it's based on my idea." The Swedish engineer who invented the zip fastener made a greater intellectual leap than that achieved by most pure academics.

Nixon's cancer program, incidentally, facilitated a lot

of good research into genetics and the structure of cells. Indeed, the overall research investment made in the twentieth century has paid off abundantly. But the payoff happens unpredictably and after a time lag that may be decades long, which is why much of science has to be funded as a public good. A fine exemplar of this point is the laser, invented in 1960. The laser applied basic ideas that Einstein had developed more than forty years earlier, but its inventors couldn't have foreseen that lasers would later be used in eye surgery and in DVD players.

Communication and assessment

Traditionally, discoveries reach public attention only after surviving peer review and being published in a scientific journal. This procedure actually dates back to the seventeenth century. In the 1660s, the Royal Society started to publish *Philosophical Transactions*, the first scientific journal, which continues to this day. Authors were enjoined to "reject all amplifications, digressions and swellings of style." This journal pioneered what is still the accepted procedure whereby scientific ideas are criticized, refined, and codified into public knowledge. Over the centuries, it published Isaac Newton's research on light, Benjamin Franklin's experiments on lightning, reports of Captain Cook's expeditions, Volta's first battery, and, more recently, many of the triumphs of twentieth-century science.

But this procedure for quality control is under increasing strain, due to competitive or commercial pressures, 24-hour media, and the greater scale and diversity of a scientific enterprise that is now widely international. (And of course scientific journals are now mainly distributed electronically rather than as paper copies.)

A conspicuous departure from traditional norms happened in 1989 when Stanley Pons and Martin Fleischmann, then at the University of Utah, claimed at a press conference to have generated nuclear power using a tabletop apparatus. If credible, this "cold fusion" fully merited the hype it aroused: it would have ranked as one of the most momentous breakthroughs since the discovery of fire. But doubts set in. Extraordinary claims demand extraordinary evidence, and in this case the evidence proved far from robust. Inconsistencies were discerned; and others failed to reproduce what Pons and Fleischmann claimed they had done. Within a year, there was a consensus that the results had been misinterpreted, though even today some believers remain.

The cold fusion claims bypassed the normal quality controls of the scientific profession, but did no great harm in the long run, except to the reputations of Pons and Fleischmann. Indeed in any similar episode today, exchanges via the Internet would have led to a consensus verdict even more quickly.

But this fiasco holds an important lesson, one that

I've already emphasized in the context of climate science. What's crucial in sifting error and validating scientific claims is *open discussion*. If Pons and Fleischmann had worked not in a university but in a lab whose mission was military, or commercially confidential, what would have happened then? If those in charge had been convinced that the scientists had stumbled on something stupendous, a massive program might have gotten under way, shielded from public scrutiny and wasting huge resources. (Indeed, just such waste has occurred, more than occasionally, in military laboratories. One example was the X-ray laser project spearheaded by Edward Teller at the Livermore Laboratory as part of President Reagan's Star Wars initiative in the 1980s.)

The imperative for openness and debate is a common thread through all the examples I've discussed. It ensures that any conclusions that emerge are robust and that science is self-correcting. Even wider discussion is needed when what's in contention is not the science itself but how new findings should be applied. Such discussions should engage all of us, as citizens, and, of course, our elected representatives. Sometimes this has happened, and constructively too. In Britain, ongoing dialogue with parliamentarians led, despite divergent ethical stances, to a generally admired legal framework on embryos and stem cells—a contrast to what happened in the United States. But Britain has had failures too: the GM crop debate was left too late, to a time when opinion was already polarized between eco-campaigners on the one side and commercial interests on the other. As a result, GM food, which has been eaten by 300 million Americans without any manifest ill effects, is severely constrained throughout the European Union.

With regard to the successful communication of science to the public, Darwin's *On the Origin of Species*, published in 1860, is an exemplar. The book was a best seller, readily accessible—even fine literature—as well as an epochal contribution to science. But that was an exception. In glaring contrast, Gregor Mendel's 1866 paper entitled "Experiments with Plant Hybrids," reporting the classic experiments on sweet peas conducted in his monastery garden, was published in an obscure journal and wasn't properly appreciated for decades. (Darwin had the journal in his private library, but the pages remained uncut. It is a scientific tragedy that he never absorbed Mendel's work, which laid the foundations for modern genetics.)

No twenty-first-century breakthroughs could be presented to general readers in such a compelling and accessible way as Darwin's ideas were; the barrier is especially high when ideas can be fully expressed only in mathematical language. Few read Einstein's original papers, even though his insights have permeated our culture. Indeed, that barrier already existed in the seventeenth century. Newton's great work, the *Principia*, highly mathematical and written in Latin, was heavy going even for his distinguished contemporaries like Halley and Hooke; certainly a general reader would have found it impenetrable, even when an English version appeared. Popularizers later distilled Newton's ideas into more accessible form—as early as 1730 a book appeared entitled *Newtonianism for Ladies*. What makes science seem forbidding is the technical vocabulary, the formulas, and so forth. Despite these impediments, the essence (albeit without the supportive arguments) can generally be conveyed by skilled communicators. It's usually necessary to eschew equations, but that by itself is not enough. The specialist jargon—and, even more, the use of familiar words (like "degenerate," "strings," or "color") in special contexts different from their everyday usage—can be baffling too.

The gulf between what is written for specialists and what is accessible to the average reader is widening. Literally millions of scientific papers are published, worldwide, each year. They are addressed to fellow specialists and typically have very few readers. This vast primary literature needs to be sifted and synthesized, otherwise not even the specialists can keep up. Moreover, professional scientists are depressingly "lay" outside their specialties—my own knowledge, such as it is, of recent biological advances comes largely from excellent popular books and journalism. Science writers and journalists do an important job, and a difficult one. I know how hard it is to explain in clear language even something I think I understand well. But journalists have the far greater challenge of assimilating and presenting topics quite new to them, often to a tight deadline; some are required to speak at short notice, without hesitation, deviation, or repetition, before a microphone or TV camera.

In Britain there is a strong tradition of science journalism. But there is an impediment: these dedicated journalists are up against the problem that few in top editorial positions have any real background in science. The editors of even the so-called highbrow press feel they cannot assume that their readers possess the level of knowledge that we might hope high school graduates would have achieved, whereas the same organs would not talk down to their readers on financial topics or on the arts pages: economic articles are often quite arcane, and the music critic would be thought to be insulting his readers if he defined a concerto or a modulation. About half of the readers of the quality press have some scientific education or are engaged in work with a technical dimension, while it is those who control the media (and those in politics) who are overwhelmingly lacking in such basic knowledge.

Science generally only earns a newspaper headline, or a place on TV bulletins, as background to a natural disaster, or health scare, rather than as a story in its own right. Scientists shouldn't complain about this any more than novelists or composers would complain that their new works don't make the news bulletins. Indeed, coverage restricted to newsworthy items—newly announced results that carry a crisp and easily summarized message—distorts and obscures the way science normally develops. Scientific ideas are better suited to documentaries and features. The terrestrial TV channels offer the largest potential audience, but commercial pressures, and concern that the viewers may channel-surf before the next advertising break, militate against the kind of extended and serious argument presented in TV classics such as Jacob Bronowski's *The Ascent of Man* and Carl Sagan's *Cosmos*. Fortunately, cable channels and the Internet open up niches for more specialized content—lectures given at universities and scientific meetings are now routinely webstreamed and archived.

The best way to ensure that one's views get through undistorted is via the written word, in articles and books. Some distinguished scientists have been successful authors but most scientists generally dislike writing, though present-day students are far more fluent (if not more literate) than my own pre–e-mail and pre-blog generation ever was. Many of the most successful writers of scientific books are interpreters and synthesizers rather than active researchers. Bill Bryson, for instance, has marvelously conveyed his zest and enthusiasm for "nearly everything" to millions.

I would derive less satisfaction from my astronomical research if I could discuss it only with professional colleagues. I enjoy sharing ideas, and the mystery and the wonder of the universe, with nonspecialists. Moreover, even when we do it badly, attempts at this kind of communication are salutory for scientists themselves, helping us to see our work in perspective. As already emphasized, researchers don't usually shoot directly for a grand goal. Unless they are geniuses (or unless they are cranks) they focus on timely, bite-sized problems because that's the methodology that pays off. But it does carry an occupational risk: we may forget that we're wearing blinkers and that our piecemeal efforts are only worthwhile insofar as they're steps toward some fundamental question.

In 1964, Arno Penzias and Robert Wilson, radio engineers at the Bell Telephone Laboratories in Holmdell, New Jersey, made, quite unexpectedly, one of the great discoveries of the twentieth century: they detected weak microwaves, filling all of space, which are actually a relic of the big bang. But Wilson afterward remarked that he was so focused on the technicalities that he didn't himself appreciate the full import of what he'd done until he read a popular description in the *New York Times*, where the background noise in his radio antenna was described as the "afterglow of creation."

Incidentally, we scientists habitually bemoan the meager public grasp of our subject—and of course all citizens need some understanding, if policy debates are to get beyond tabloid slogans. But maybe we protest too much. On the contrary, we should, perhaps, be gratified and surprised that there's wide interest in such remote topics as dinosaurs, the Large Hadron Collider in Geneva, or alien life. It is indeed sad if some citizens can't distinguish a proton from a protein; but equally so if they are ignorant of their nation's history, or are unable to find Korea or Syria on a map—and many people can't.

Misperceptions about Darwin or dinosaurs are an intellectual loss, but no more. In the *medical* arena, however, they could be a matter of life and death. Hope can be cruelly raised by claims of miracle cures; exaggerated scares can distort health-care choices. When reporting a particular viewpoint, journalists should clarify whether it is widely supported, or whether it is contested by 99 percent of specialists. The latter was the case when a doctor claimed that the MMR vaccine (offering combined protection against measles, mumps, and rubella) could induce autism in small children—a claim that was later discredited.

Noisy controversy need not signify evenly balanced arguments. Of course, the establishment is sometimes routed and a maverick vindicated. We all enjoy seeing this happen, but such instances are rarer than is commonly supposed. The best scientific journalists and bloggers are plugged into an extensive network that should enable them to calibrate the quality of novel claims and the reliability of sources.

But what about ideas beyond the fringe? Here there's less scope for debate—the two sides do not share the same methods or play by the same evidence-based rules; as an

astronomer, I've not found it fruitful to have much dialogue with astrologers nor creationists. We shouldn't let a
craving for certainty—for the easy answers that science can
seldom offer—drive us toward the illusory comfort and
reassurance that such concepts offer.

Scientists should expect media scrutiny. Their expertise
is crucial in areas that fascinate us and matter to us all. And
they shouldn't be bashful in proclaiming the overall promise that science offers—it's an unending quest to understand nature, and essential for our survival.

Scientists as campaigners and concerned citizens

I'll end, as I began, with a flashback, this time to World
War II. The scientific community was then engaged in the
war effort, most monumentally in the Manhattan Project
that led to the development of the first atomic bomb, but
also in radar, operations research, and code-breaking. Most
scientists returned with relief to peacetime academic pur-
suits but for some, especially those who had helped build
the bomb, the ivory tower was no sanctuary. They continued not just as scientists but as engaged citizens, promoting
efforts to control the power they had helped to unleash.

Among them was Joseph Rotblat, a physicist of Polish
origin who went to Los Alamos as part of the UK scientific
contingent. In his mind there was only one justification for
the bomb project: to ensure that Hitler didn't get one first

and hold the Allies to ransom. As soon as this ceased to be a credible risk, Rotblat left the Manhattan Project, the only scientist to do so at that juncture. He returned to England, became a professor of medical physics, an expert on the effects of radiation, and a compelling and outspoken campaigner. In 1955, he met Bertrand Russell and encouraged him to prepare a manifesto stressing the extreme gravity of the nuclear peril. Rotblat got Einstein to sign, too; it was Einstein's last public act—he died a week later. This Russell-Einstein manifesto was then signed by nine other eminent scientists from around the world. It led to the initiation of the Pugwash Conferences—so-called after the village in Nova Scotia where the inaugural conference was held. These meetings, which continue to this day, helped to sustain a dialogue between scientists in Russia and the West throughout the Cold War. Such contacts eased the path for the Partial Test Ban Treaty of 1963 and the later Anti-Ballistic Missile Treaty. When the Pugwash Conferences were recognized by the 1995 Nobel Peace Prize, half the award went to the Pugwash organization and half to Rotblat personally, as their prime mover and untiring inspiration.

The goal of Rotblat's crusade was to rid the world completely of nuclear weapons, an aim that was widely derided as the product of woolly idealism. But this goal gained broader establishment support over the years and in 2006

the Gang of Four—George Shultz, Sam Nunn, William
Perry, and Henry Kissinger—espoused a similar cause.
They have been supplemented by European gatherings of
senior politicians, including former ministers of defense.
More importantly, President Obama has reactivated the
disarmament agenda by persuading the United States Sen-
ate to ratify the SALT agreement, and—in, for instance, his
inspirational speech in Prague in 2010—has set zero as an
ultimate goal.

Few of the generation with senior involvement in World
War II are alive today. In the United States, they have been
followed by an impressive cohort of scientists—people
from succeeding generations who have done a spell in government, or in high-tech industry, and who serve regularly
as consultants to the Pentagon or on advisory committees.
But in my own country, Britain, there are depressingly
few younger scientists who can match the credentials and
expertise of their US counterparts in providing independent expertise. The reasons for this transatlantic asymme-
try aren't hard to find. In the United States, senior staff
shuffle between government jobs and posts in, for instance,
the Brookings Institution whenever the administration
changes. There are always some who are "out" rather than
"in." Britain, in contrast, doesn't have a revolving-door
system; government service is still generally a lifetime
career. For this reason, and because secrecy is more perva-

sive, discussions of defense issues tend to be restricted to a closed official world.

The atomic scientists of the World War II generation were an elite group—the alchemists of their time, possessors of secret knowledge—and independent British scientists cannot aspire to the wisdom and expertise of that battle-hardened generation. But defense and arms control are a diminishing part of the agenda for today's citizen scientists: the agenda is now far wider and more complex—and the issues span all of the sciences. They are far more open and often global. There is less demarcation between experts and laypersons; campaigners and bloggers enrich the debate. But professionals have special obligations to engage and men like Rotblat were inspiring exemplars. You would be a poor parent if you didn't care about what happened to your children in adulthood, even though you may have little control over it. Likewise, scientists, whatever their expertise, shouldn't be indifferent to the fruits of their ideas. Their influence may be limited, but they should try to foster benign spin-offs, commercial or otherwise. They should resist, so far as they can, dubious or threatening applications of their work and alert the public and politicians to perceived dangers.

Unprecedented pressures confront the world, but there are unprecedented prospects too. The benefits of globalization must be fairly shared. There's a widening gap between what science allows us to do and what it's prudent or ethical actually to do—there are doors that science could open but which are best left closed. Everyone should engage with these choices, but their efforts must be leveraged by scientific citizens—engaging, from all political perspectives, with the media, and with a public attuned to the scope and the limit of science.

2

SURVIVING THE CENTURY

As an astronomer, I sometimes get mistaken for an astrol-oger—but I cast no horoscopes and have no crystal ball. The past record of scientific forecasters is, in fact, dismal. When Alexander Graham Bell invented the telephone, he enthused that "some day, every town in America will have one." The great physicist Lord Rutherford averred that nuclear energy was moonshine; Thomas Watson, founder of IBM, said, "I think there is a world market for maybe five computers"; and one of my predecessors as England's Astronomer Royal said that space travel was "utter bilge." I won't add to this inglorious roll call. Instead, let us focus on a key question: How can our scientific capabilities be deployed to ease rather than aggravate the tensions that the world will confront in the coming decades?

Our lives today are molded by three innovations that

gestated in the 1950s, but whose pervasive impact certainly wasn't then foreseen. Indeed, forecasters generally underestimate long-term changes, even when overestimating short-term ones. It was in 1958 that Jack Kilby and Robert Noyce built the first integrated circuit, the precursor of today's ubiquitous silicon chip—perhaps the most transformative single invention of the last century. It has spawned the worldwide reach of mobile phones and Internet, promoting economic growth while itself being sparing of energy and resources. In the same decade James Watson and Francis Crick discovered the bedrock mechanism of heredity—the famous double helix. This launched the science of molecular biology, opening profound prospects whose main impact still lies ahead. Ten years ago, the first draft of the human genome was decoded. It was a huge international project, acclaimed by President Clinton and Prime Minister Blair at a special press conference, and the cost was around $3 billion. Today, genome sequencing—the "read out" of our genetic inheritance—is becoming a routine technique that costs only a few thousand dollars.

And there's a third technology that emerged during this period: space. It's over fifty years since the launch of *Sputnik*, an event that led President Kennedy to inaugurate the Apollo program to land men on the Moon. Kennedy's prime motive was of course superpower rivalry; cynics could deride it as a stunt but it was undoubtedly an extraordinary technical triumph. And *Apollo* had an inspirational legacy too. Distant images of Earth—its delicate biosphere of clouds, land, and oceans contrasting with the sterile moonscape where the astronauts left their footprints—have, ever since the 1960s, been iconic for environmentalists.

But of course there was always a dark side to space. Rockets were primarily developed in order to carry nuclear weapons, and those weapons were themselves the outcome of the World War II Manhattan Project, which inaugurated the nuclear age and was even more intense and focused than the Apollo program. We lived, throughout the Cold War, under a threat of nuclear catastrophe that could have shattered the fabric of civilization, a threat especially acute at the time of the Cuba crisis in 1962. It wasn't until he'd long retired that Robert McNamara, then US Secretary of Defense, spoke frankly about the events in which he'd been so deeply implicated. In his confessional documentary film *The Fog of War*, he said, "We came within a hair's-breadth of nuclear war without realizing it. It's no credit to us that we escaped—Krushchev and Kennedy were lucky as well as wise." Indeed on several occasions during the Cold War the superpowers could have stumbled toward armageddon.

The threat of *global* nuclear annihilation involving tens of thousands of bombs is, thankfully, in abeyance, but this prospect hasn't gone for good: we can't rule out, by mid-century, a global political realignment leading to a standoff between new superpowers that could be handled less well or less luckily than was the Cuban missile crisis. And the risk that smaller nuclear arsenals proliferate, and are used in a regional context, is higher than it ever was. Moreover, al-Qaida-style terrorists might someday acquire a nuclear weapon and willingly detonate it in a city, killing tens of thousands along with themselves.

The nuclear age inaugurated an era when humans could threaten the entire Earth's future. We'll never be completely rid of the nuclear threat—H-bombs can't be disinvented—but the twenty-first century confronts us with grave new perils. They may not threaten a sudden worldwide catastrophe, but they are, in aggregate, worrying and challenging. Some will be consequences of new technologies that we can't yet envisage—any more than Rutherford could have predicted thermonuclear weapons.

World population trends

There's one trend that we can predict with confidence. There will, by midcentury, be far more people on the Earth than there are today. Fifty years ago, world population was below 3 billion. It has more than doubled since then and reached 7 billion in 2011; the projections for 2050 range between 8.5 and 10 billion, the growth being mainly in the developing world. More than 50 percent of the world's population now live in cities, and this proportion is growing.

The majority of the world's people live in countries

where fertility has fallen below the replacement level of about 2.1 births per woman. The falls over the last twenty-five years have been dramatic (by 50 percent in Brazil, and more than 70 percent in Iran). The current rates range from 7.1 in Niger to 1.2 in South Korea; the European average is about 1.4. This so-called demographic transition is a consequence of declining infant mortality, availability of contraceptive advice, and women's education, among other things. However, more than half of those in the developing world who are alive today are less than twenty-five years old—and with rising life expectancy. That's why a continuing population rise until midcentury seems almost inevitable. If the demographic transition quickly extended to all countries, then the global population could gradually decline after 2050. But numbers are rising fast in India, whose population is projected to overtake China's by 2030 and could exceed 1.6 billion by 2050. Population projections for Africa are also rising. A hundred years ago, Ethiopia's population was 5 million. That figure is now about 80 million and is predicted to almost double by 2050. The populations of both Sudan and Uganda are also estimated to more than double by midcentury, putting increasing pressure on the water resources of the Nile basin. In total, estimates suggest that there could be a billion more people in Africa in 2050 than there are today.

But the trends beyond 2050 will depend on what people now in their teens and twenties decide about the number and spacing of their children. In Africa there are around 200 million women who are denied such a choice. Enhancing the life chances of Africa's poorest people—by providing clean water, primary education, and other basics—should be a humanitarian imperative. But it would seem also a precondition for achieving throughout that continent the demographic transition that has occurred elsewhere. Failure to achieve this would be a failure of governance—the resources required are modest—and a continental-scale tragedy that would also trigger massive migratory pressures. (An earlier Reith Lecturer, Jeffrey Sachs, pointed out that the resources needed to achieve the UN's millennium goals, and thereby enhance the lives of the world's "bottom billion," were less than those possessed by the thousand hyperwealthy individuals in the world.)

To feed Africa's people, farming productivity must be enhanced (but without degrading the soil or ecology, as is now happening) using modern agricultural methods, among which genetic modification could be one. And water shortages, perhaps aggravated by climate change, must be contended with. To produce a kilogram of vegetables with present methods takes 2,000 liters of water; a kilogram of beef takes 15,000 liters. But as well as biological advances, modern engineering practices must be adopted to conserve water, reduce food waste, and so on.

It is not possible to specify an optimum population for the world. This is because we can't confidently conceive what people's lifestyles, diet, travel patterns, and energy needs will be beyond 2050. The world couldn't sustain anywhere near its present population if everyone lived like present-day Americans, profligate of energy and resources. On the other hand, more than 10 billion people could live sustainably, with a high quality of life, if all adopted a vegetarian diet, traveling little but interacting via super-Internet and virtual reality (though this particular scenario is plainly not probable, nor, necessarily, attractive). So it's naive to quote a single unqualified headline figure for the world's carrying capacity or optimum population.

Two messages, in particular, should surely be proclaimed more widely. First, enhanced education and empowerment of women within this decade—surely a benign priority in itself—would reduce fertility rates in the poorest nations, and could thereby reduce the projected world population beyond 2050 by as much as a billion. Second, the higher the post-2050 population becomes, the greater the pressures on resources—especially if the developing world, where most of the growth will be, narrows its gap with the developed world in its per capita consumption of energy and other resources.

Over 200 years ago, Thomas Malthus famously argued that populations would rise until limited by food shortages. His gloomy prognosis has been forestalled—despite a sevenfold rise in population since his time—by advancing technology and the green revolution. He could one day be tragically vindicated. But this need not happen. There is real potential, in Africa and elsewhere, for sustainably enhancing food production so that the larger and much demanding population expected in 2050 can be fed—but only if current knowledge and techniques are efficiently and appropriately applied.

It is important to highlight global population because the topic seems currently underdiscussed. That is because doom-laden forecasts made in the past have proved off the mark, and because it is deemed by some a taboo subject— tainted by association with eugenics in the 1920s and 1930s, with Indian policies under Indira Gandhi, and, more recently, with China's effective but hard-line one-child policy.

Responses to potential climate change

Another firm prediction about the post-2050 world is that, as well as being more crowded, it will on average be warmer. Human actions—mainly the burning of fossil fuels—have already raised the carbon dioxide concentration higher than it's ever been in the last half million years. The graph of carbon dioxide concentration, the so-called Keeling Curve, shows a steady rise over the last fifty years—a mere geological instant—superimposed on the readily detectable annual oscillations caused by the seasonal growth and decay of vegetation in the Northern Hemisphere. These beautiful and precise measurements are entirely uncontroversial. Moreover, according to business-as-usual scenarios where we remain as dependent as today on fossil fuels, the concentration could reach twice the preindustrial level within fifty years, and go on rising. This much is uncontroversial too. Nor is there significant doubt that CO_2 is a greenhouse gas, and that the higher its concentration rises, the greater the warming—and, more important still, the greater the chance of triggering something grave and irreversible: rising sea levels due to the melting of Greenland's ice cap; runaway greenhouse warming due to release of methane in the tundra; and so on.

There is, as mentioned in Chapter 1, still substantial uncertainty about just how sensitive the temperature is to the CO_2 level. The climate models can, however, assess the likelihood of a range of temperature rises. It is the high-end tail of the probability distribution that should worry us most: the small probability of a really drastic climatic shift. Climate scientists now aim to refine their calculations, and to address questions like: Where will the flood risks be concentrated? What parts of Africa will suffer severest drought? Where will the worst hurricanes strike? Will extreme weather events become more frequent? The headline figures routinely quoted—2-, 3-, or 5-degree rises in the mean global temperature—might seem too small to fuss about, but two comments should put this into perspective. First, even in the depth of the last ice age the mean temperature was lower by just 5 degrees. Second, it's important to realize that the rise won't be uniform: the land warms more than the sea, and high latitudes more than low. Quoting a single figure glosses over shifts in global weather patterns that will be more drastic in some regions than in others (and indeed may cause some regions to cool rather than warm). A mean global rise of 4 degrees could lead to a warming of 10 degrees centigrade (18 degrees Fahrenheit) in western and southern Africa. Indeed the worst effects of any warming may be in Africa and Bangladesh, which have contributed least to the emission. Rising carbon dioxide could induce relatively sudden flips rather than just gradual changes.

The science of climate change is intricate, but it's straightforward compared to the economics and politics. The economist Nicholas Stern, in his influential review "The Economics of Climate Change," written for the British government in 2006, averred that a response to global warming needs "all the economics you ever learnt, and some more. It's a market failure on a colossal scale." It poses a unique political challenge for two reasons. First, the effect is nonlocalized: CO_2 emissions from the United States have no more effect there than they do in Australia, and vice versa. That means that any credible regime whereby the polluter pays has to be broadly international. Second, there are long time lags; it takes decades for the oceans to adjust to a new equilibrium, and centuries for ice sheets to melt completely. Even though anthropogenic climate change is already perceptible, the main downsides of global warming lie a century or more in the future. Concepts of intergenerational justice then come into play: How should we rate the rights and interests of future generations compared to our own? What discount rate should we apply?

The declared political goal has been to halve global carbon dioxide emissions by 2050. On the basis of the best current modeling, this is the reduction needed in order to bring below 50 percent the probability that the mean temperature will rise by more than 2 degrees. This target corresponds to a ration of 2 tons of carbon dioxide per year for each person on the planet. For comparison, the current US level is 20 tons per person, per year, the European figure is about 10, the Chinese level is already 5.5, and the Indian is 1.5. This target must be achieved without stifling economic growth in the developing world, where the emissions in the short term are bound to rise, so it's the richer countries that must take the lead in making cuts. (In particular, nothing should take priority over quick action to bring electricity to the poorest, whose meager power supply, from the burning of cow dung or wood, is hazardous to health. Deaths from indoor household fuel pollution, worldwide, run at 1.6 million per year—as many as from malaria. Two billion lives could be transformed without adding more than 1 percent to current emissions.)

Success in halving global carbon emissions would be a momentous achievement—one where all nations acted together, in the interests of a future beyond the normal political horizon. The meager progress in Copenhagen in December 2009 led to a pessimism that was only partly allayed by the outcome of the Cancun meeting a year later. On the other hand, odd though this may sound, the political response to the 2009 financial crisis may offer encouragement. Who would have thought three years ago that the world's financial system would have been so transformed that big banks were nationalized? Likewise, we need coordinated outside-the-box action to avoid serious risk of a long-term energy crisis.

There is, incidentally, at least one precedent for long-term altruism. In discussing the safe disposal of nuclear waste, policy makers talk with a straight face about what might happen more than 10,000 years from now, thereby implicitly applying zero discount to the cost of future hazards or benefits. To concern ourselves with such a remote "posthuman" era might seem bizarre, but all of us can surely empathize at least a century ahead. Especially in Europe, we're mindful of the heritage we owe to centuries past and history will judge us harshly if we discount too heavily what might happen when our grandchildren grow old.

The European Union (EU) is pursuing progressive policies aimed at reducing the carbon emissions of its member states. Yet whatever the EU and the rest of the world do, if the United States and China do not alter their current policies there is little or no hope of meeting the targets: these two giants between them contribute over 40 percent of total global emissions. Despite the pledge made by President Obama at his inauguration that containing climate change was a priority goal, political paralysis seemingly prevails at the federal level, though there are positive initiatives at the level of local communities, third-sector organizations, cities, and states. China's leaders (many of whom are trained engineers) seem aware of their nation's vulnerability to climate change and are investing in renewable technologies and nuclear power on a substantial scale. They emphasize the need to improve energy efficiency, but the rapid economic growth (and dependence on coal-fired power stations) is generating emissions that are still increasing year by year, albeit more slowly than the GNP. Whatever targets are set, they won't be met without a transition to a lifestyle dependent on clean and efficient energy technology. Enlightened business leaders recognize that the actions needed to abate the threat of climate change will create manifold new economic opportunities.

The world spends more than $5 trillion a year on energy and its infrastructure. But currently far too little is invested in developing techniques for economizing on energy, storing it, and generating it by low-carbon methods. Certainly the major utilities are spending very little: according to the journalist Tom Friedman, US energy utilities spend less on R&D than the American pet-food industry does. The main US investments are in small start-up companies, especially in solar energy. (In Britain, meanwhile, R&D within the energy industries is now expanding, but has yet to fully attain the pre-privatization level of the late 1980s.) There's a glaring contrast here with the fields of health and medicine, where the worldwide R&D expenditures are disproportionately higher. The clean energy challenge deserves a priority and commitment akin to the Manhattan Project or the Apollo Moon-landing. Indeed it would be hard to think of anything more likely to enthuse young people toward careers in engineering than a firmly proclaimed priority to develop clean energy for the developing and the developed world.

My own country, Britain, is so small that its stance may seem of marginal import: our carbon emissions constitute only 1 or 2 percent of the problem. But we have tried to exert leverage in two respects. We gained influence because of Tony Blair's efforts at the Gleneagles G8 Summit, which he hosted in 2005, and have already enshrined in the Climate Change Act a commitment to cut our own emissions by 80 percent over the next forty years. It's important to give credit to several politicians, of all parties, who worked hard to keep these issues high on the agenda even though long-term altruism is plainly not a vote-winner; the Conservative-led coalition currently in power has not yet backtracked.

The United States and Europe should both prioritize the technologies needed for a low-carbon economy.
There's a need to ensure our own energy, but, beyond that
imperative, it's in our interest not to fall behind the Chinese in developing clean energy technologies for which the
demand will be worldwide.

Wave or tidal energy is a niche market, but it's one where
Britain has a competitive advantage. This island nation has
the geography—capes round its coast with fast-flowing
tidal currents—and also expertise in marine technology
spun off from North Sea oil and gas projects. There is a
long-studied scheme to get 7 percent of our electric power
by tapping the exceptional (40-foot) tidal range in the estu-
ary of the River Severn by a 10-mile-long barrage.

What about biofuels? There's been ambivalence about
them because they compete for land use with food-growing
and forests, but in the long run GM techniques may lead
to novel developments: bugs that break down cellulose,
or marine algae that convert solar energy directly into
fuel. Another need is for improved energy storage. Steven
Chu, the Nobel Prize–winning physicist whom President
Obama appointed as Energy Secretary, has given priority
to improving batteries—for electric cars, and to comple-
ment unsteady power sources such as sun and wind.

As to the role of nuclear power, I myself would favor
Britain and the United States having at least a replacement
generation of power stations. But risks can be catastrophic

and the nuclear nonproliferation regime is fragile. One can't be relaxed about a worldwide program of nuclear power unless internationally regulated fuel banks are established to provide enriched uranium and remove and store the waste—and unless there is a strictly enforced safety code to guard against risks analogous to those from poorly maintained "third-world airlines." Despite the ambivalence about widespread nuclear energy, it's surely worthwhile to boost R&D into fourth-generation reactors, which could be more flexible in size, and safer. The industry has been relatively dormant for the last twenty years, and current designs date back to the 1960s.

And of course nuclear fusion, the process that powers the Sun, still beckons as an inexhaustible source of energy. Attempts to harness this power have been pursued ever since the 1950s, but the history here is of receding horizons: commercial fusion power is still at least thirty years away. The challenge is to use magnetic forces to confine gas at a temperature of millions of degrees—as hot as the center of the Sun. Despite its cost, the potential payoff is so great that it is surely worth continuing to develop experiments and prototypes. The largest such effort is the International Thermonuclear Experimental Reactor, internationally funded and based in France; similar projects are being pursued in Korea and elsewhere. (These all involve magnetic confinement of ultrahot gas. An alternative concept, whereby tiny deuterium pellets are imploded and heated by converging beams from immense lasers, is being pursued at the Livermore Laboratory in California, but this seems primarily a defense project to provide lab-scale substitutes for H-bomb tests, where the promise of controlled fusion power is a political fig leaf.)

An attractive long-term option is solar energy—huge collectors in the desert could generate power that's distributed via a continent-wide smart grid. Achieving this would require vision, commitment, and public-private investment on the same scale as the building of the railways in the nineteenth century. Indeed, any transformation of the world's energy and transport infrastructure is a massive project that would take several decades. (And to match the United States and China the EU needs a more coordinated and less labyrinthine organizational structure. Henry Kissinger once famously complained, "When I want to speak to Europe, whom do I call?" Pan-European decisions are needed on more and more issues.) Even those who favor renewable energy in principle often oppose specific proposals because of their environmental or aesthetic impact. But we need to remember that all the renewables—wind, tidal, solar, or biofuels—have downsides so some such impact is unavoidable. So of course does nuclear power, though the aggregate depends on how much energy per capita is needed for the lifestyle that we want. And of course alternative energy sources would supersede coal mines and oil rigs, of whose environmental and human risks we are all too aware already.

Many of us still hope that our civilization can segue toward a low-carbon future and a lower population, and can achieve this transition without trauma and disaster. But that will need determined action by governments, urgently implemented; and such urgency won't be achieved unless sustained campaigning can transform public attitudes and lifestyles. Of course, no politician will gain much resonance by advocating a bare-bones approach that entails unwelcome lifestyle changes. The priority for all developed countries should be to implement measures that actually save money—by using energy more efficiently, insulating buildings better—and to incentivise new clean technologies so that, as fossil fuel prices rise, a transition to clean energy is less costly. But what is *very* important is to prioritize the development of those new energy sources, be they wind, tides, solar, or nuclear. There may be need for a new international body, along the lines of the World Health Organization or the International Atomic Energy Agency, to monitor fossil fuel use and facilitate the transition to clean energy.

The strongest motive for urgent action is that the worst-case climatic scenarios predict really serious stresses. Even within fifty years, we could face serious climatic stresses if effective mitigation is too-long postponed and the warming rate is at the upper end of the predicted range.

In twenty years, we will know—from firmer science, improved computer modeling, and also from a longer time span of data on actual climatic trends—whether the feedback from water vapor and clouds strongly amplifies the effect of CO_2 itself in creating a greenhouse effect. If so, and if the world consequently seems on a rapidly warming trajectory because international efforts to reduce emissions haven't been successful, there may be a pressure for panic measures. These would have to involve a plan B—being fatalistic about continuing dependence on fossil fuels, but combating its effects by some form of geoengineering. One option is to counteract the greenhouse warming by, for instance, putting reflecting aerosols in the upper atmosphere or even vast sunshades in space. The political problems of such geoengineering may be overwhelming. Not all nations would want to turn down the thermostat equally, and there could be unintended side effects. Moreover, the warming would return with a vengeance if the countermeasures were ever discontinued; and other consequences of rising CO_2 (especially the deleterious effects of ocean acidification) would be unchecked. An alternative strategy, which currently seems less practicable, would involve direct extraction of carbon from the atmosphere, either by deploying, on vast scales, the principles used by scrubbers to purify air in submarines, or else by growing trees and "fixing" the carbon they absorb as charcoal. This approach would be politically more acceptable: we'd essentially just be undoing the unwitting geoengineering we've done by burning fossil fuels.

It seems prudent at least to study geoengineering, to clarify which options make sense and perhaps counter undue optimism about a technical quick fix of our climate. However, it already seems clear that it would be feasible and affordable to throw enough material into the stratosphere to change the world's climate—indeed what is scary is that this capacity might be within the resources of a single nation, or even a corporation or plutocratic individual. Very elaborate climatic modeling would be needed in order to calculate the regional impacts of such an intervention. That is why it is crucial to sort out the complex governance issues raised by geoengineering and to do this well before there is any chance that urgent pressures for action will build up.

Other vulnerabilities

Energy security, food supplies, and climate change are the prime long-term "threats without enemies" that confront us, all aggravated by rising populations. But there are others. For instance, rapid changes in land use may jeopardize whole ecosystems. There have been five great extinctions in the geological past; human actions are causing a sixth. The extinction rate is a hundred, even a thousand, times higher than normal. We are destroying the book of life before we have read it. To quote E. O. Wilson, one of the world's most distinguished ecologists, certainly its most eloquent: "At the heart of the environmentalist world view is the conviction that human physical and spiritual health depends on the planet Earth . . . Natural ecosystems— forests, coral reefs, marine blue waters—maintain the world as we would wish it to be maintained. Our body and our mind evolved to live in this particular planetary environment and no other."

Biodiversity is often proclaimed as a crucial component of human well-being and economic growth. It manifestly is: we're clearly harmed if fish stocks dwindle to extinction; there are plants in the rain forest whose gene pool might be useful to us. But for Wilson these instrumental—and anthropocentric—arguments aren't the only compelling ones. For him, preserving the richness of our biosphere has value in its own right, over and above what it means to us humans.

So far my focus has been on threats that we're *collectively* imposing on the biosphere. But it is important also to highlight a new type of vulnerability that could stem from empowerment of individuals or small groups by fast-developing technologies.

Almost all innovations entail new risks. Most surgical procedures, even if now routine, were often fatal when pioneered; and, in the early days of steam, people died when poorly designed boilers exploded. But something has changed. Hitherto, most of the risks caused by humans have been localized and limited. If a boiler explodes it is horrible, but there is a limit to just how horrible. But there are new hazards whose consequences could be so wide-spread that even a tiny probability is disquieting. Global society is precariously dependent on elaborate networks—electricity grids, air traffic control, international finance, just-in-time delivery, and so forth. It's crucial to ensure maximal resilience of all such systems, and maximum security against sabotage, at a time when concern about cyber attack, by criminals or by hostile nations, is rising sharply. Otherwise the manifest benefits of those systems could be outweighed by catastrophic (albeit rare) break-downs cascading through them.

There are other potential vulnerabilities. It's becom-ing feasible, for instance, to stitch together long strands of DNA, and thereby construct from scratch the blueprint of an organism. The potential for medicine and agriculture is huge, but there are risks too. Already the genomes for some viruses—polio, Spanish flu, and SARS—have been synthesized. Expertise in such techniques will become widespread, posing a manifest risk of bioerror or bioter-ror. In the early days of recombinant DNA (gene splicing) in the 1970s, there was concern about unintended conse-quences, and a moratorium was imposed after a conference of experts held at Asilomar, California. This moratorium soon came to seem unduly cautious, but that doesn't mean that it was unwise at the time, since the level of risk was then genuinely uncertain. It showed that an international group of leading scientists could agree to a self-denying ordinance, and influence the research community powerfully enough to ensure that it was implemented. There have recently been moves to control the still more powerful techniques of synthetic biology that can create organisms from the ground up by synthesizing a new genome bit by bit. A voluntary consensus is harder to achieve today: the academic community is far larger, and competition (enhanced by commercial pressures) is more intense.

We're kidding ourselves if we think that those with technical expertise will all be balanced and rational: expertise can be allied with fanaticism—not just the types of fundamentalism with which we are currently familiar, but that exemplified by some New Age cults, extreme ecofreaks, violent animal rights campaigners, and the like. And there will be individual weirdos, with the mindset of those who now unleash computer viruses—the mindset of arsonists. The global village will have its village idiots. In a future era of vast individual empowerment, where even one malign or foolish act could be too many, how can our open society be safeguarded? Perhaps our society will make a shift toward more intrusion and less privacy. (Indeed, the rash abandon with which people put their intimate details on Facebook, and our acquiescence in ubiquitous CCTV, suggests that such a shift would meet surprisingly little resistance.) Or will there be pressures to constrain diversity and individualism? These may become serious issues.

Some years ago I wrote a short book on the theme of

this chapter, entitled *Our Final Century?* My British publishers deleted the question mark; the American publishers
changed the title to *Our Final Hour* (the US public seeks
instant [dis]gratification). I conjectured that, taking all risks
into account, there was only a 50 percent chance that we
would get through to 2100 without a disastrous setback.
This seemed a depressing conclusion. However, I have
been surprised by how many of my colleagues thought a
catastrophe was even more likely than I did, and so con-
sidered me an optimist. I am actually an optimist—at least
a *techno*-optimist. Over the last decade, we've experienced
astonishing advances in communication and in access to
information. Our lives have been hugely enriched by con-
sumer electronics and web-based services that we would
willingly pay far more for, and which surpass any expecta-
tions we had a decade ago. And the impact on the developing world has been dramatic: there are more mobile phones
than toilets in India. Mobile phones have penetrated Africa
too, offering rural farmers access to market information
that prevents them from being ripped off by traders, and
enabling money transfers. There is now great demand for
low-power solar generators in order to charge them up. (In
his 2010 BBC series *The World in 100 Objects*, Neil Mac-
Gregor, Director of the British Museum, chose a mobile
phone and its charger as the 100th object—emblematic of
the transformational power of optimally applied science
in impoverished regions.) Broadband Internet, soon to

achieve worldwide reach, should further stimulate education and the adoption of modern health care, agriculture, and technology.

Recent history augurs well for the transformational power of optimally applied science; and we can hope that health and agriculture surge ahead likewise. There seems no scientific impediment to achieving a sustainable world beyond 2050, in which the developing countries have narrowed the gap with the developed, and all benefit from further advances that could have as great and benign an impact as information technology has had in the last decade. But the intractable politics and sociology—the gap between potentialities and what actually happens—engender pessimism. Will richer countries recognize that it's in their self-interest for the developing world to prosper, sharing fully in the benefits of globalization? Can nations sustain effective but nonrepressive governance in the face of threats from small groups with high-tech expertise? And can the focus of our sympathies become more broadly international? And, above all, can our institutions prioritize projects that are long-term in political perspective even if a mere instant in the history of our planet?

A cosmic perspective

I'll conclude with a cosmic vignette. Suppose some aliens had been watching our planet from afar for its entire history. What would they have seen? Over nearly all that immense time, 45 million centuries, Earth's appearance would have altered very gradually. Continents drifted; the ice cover waxed and waned; successive species emerged, evolved, and became extinct. But in just a tiny sliver of the Earth's history—the last one-millionth part—patterns of vegetation altered at an accelerating rate. This signaled the advent of agriculture and the growing impact of humans.

Then, in just one century, came other changes. The amount of carbon dioxide in the air began to rise anomalously fast. The planet became an intense emitter of radio waves (the output from TV, cell phone, and radar transmissions). And something else unprecedented happened: small projectiles launched from the planet's surface escaped the biosphere completely. Some were propelled into orbits around the Earth; some journeyed to the Moon and planets. If they understood astrophysics, the aliens could predict that the biosphere would face doom in a few billion years when the Sun flares up and dies. But could they have predicted this sudden "fever" less than halfway through the Earth's life?

And if they continued to keep watch, what might these hypothetical aliens witness in the next hundred years—in this unique century? Will a final spasm be followed by silence? Or will the planet itself stabilize? And will some of the objects launched from the Earth spawn new oases of life elsewhere? The answer depends on how the challenges

I've addressed can be met. Twenty-first-century science, if optimally applied, could offer immense benefits to the developing and the developed worlds—but it will present new threats. To confront these successfully—and to avoid foreclosing humanity's long-term potential—is the political and social challenge for the coming decades.

3

WHAT WE'LL
NEVER KNOW

While struggling to prepare my lectures for BBC radio, I had a fantasy. Suppose I had a time machine. I could fast forward into the future, turn on the radio, listen to the broadcast version, take notes—and then reverse back to the present and start writing. Well, there was plainly no such quick fix—but could there ever be?

Arthur C. Clarke noted that any sufficiently advanced technology is indistinguishable from magic. We can't now envision what artifacts might exist centuries hence, any more than a Roman could imagine today's SatNav and mobile phones. Nevertheless, a physicist would confidently assert that time machines will remain forever fiction. That's because changing the past would lead to paradoxes—infanticide would violate logic as well as ethics if the victim was your grandmother. So, what is the demarcation between concepts that seem crazy now but might be realized eventually and things that are forever impossible? Are there limits to how much we can ever predict? Are there scientific problems that will forever baffle us—phenomena that simply transcend human understanding?

Einstein averred that "the most incomprehensible thing about the universe is that it is comprehensible." He was right to be astonished. Our minds evolved to cope with life on the African savannah, but they can also comprehend the microworld of atoms and the vastness of the cosmos. We marvel at the fact that the universe isn't anarchic—that atoms obey the same laws in distant galaxies as in the lab. Our cosmic horizons have vastly enlarged. Our Sun is one of a hundred billion stars in our galaxy, which is itself one of many billion galaxies in range of our telescopes. And this entire panorama emerged from a hot, dense beginning nearly 14 billion years ago. Some inferences about the early universe are as evidence-based as anything a geologist might tell you about the history of our Earth; we can make confident and precise statements about how hot and dense things were in the first few seconds of our universe's expansion, even just a microsecond after the big bang. But, as always in science, each advance brings into focus some new questions that couldn't previously have even been posed. We now confront the mystery of the very beginning (if indeed there was one). It's a mystery because, right back in the first tiny fraction of a second, conditions would have been far hotter and denser than we can simulate in the lab. We don't know the physical laws that then prevailed, so we lose any foothold in experiment.

Einstein himself made one of the biggest advances in our comprehension. More than 200 years before him, Isaac Newton had shown that the gravity that makes apples fall is the same force that holds planets in their orbits. Einstein went much further. He didn't prove Newton wrong— Newton's mathematics is good enough to program space flights to distant planets—but he transcended Newton by offering insights into gravity that made it seem more natural and linked it to the nature of space and time, and the universe itself.

The other great twentieth-century intellectual revolution, quantum mechanics, tells us, completely counter to all intuition, that on the atomic scale nature is intrinsically fuzzy; nonetheless, atoms behave in precise mathematical ways when they emit and absorb light, or link together to make molecules. Quantum theory is the basis for much of modern technology: it is vindicated every time you take a digital photograph, surf the Internet, or use any gadget—a DVD player, or a supermarket bar code—that involves a laser. Quantum mechanics fully matched the importance of Einstein's work, but was a collective rather than individual effort.

General relativity and quantum theory are the twin pillars of twentieth-century physics, but at the deepest level they contradict each other—they haven't yet been meshed into a single unified theory. In most contexts, this does not impede us because their domains of relevance do not overlap. Astronomers can ignore quantum fuzziness when calculating the motions of planets and stars. Conversely, chemists can safely ignore gravitational forces between individual atoms in a molecule because these are nearly forty powers of ten feebler than electrical forces. But during the very earliest instants after the big bang, when *everything* was squeezed smaller than a single atom, quantum fluctuations could shake the entire universe.

Mysteries of the cosmos and microworld

To confront the overwhelming mystery of what banged and why it banged, Einstein's theory isn't enough because it treats space and time as smooth and continuous. Success will require new insights into what might seem the simplest entity of all: "mere" empty space. We know that no material can be chopped into arbitrarily small pieces: eventually you get down to discrete atoms. Likewise, even space and time can't be divided up indefinitely. There are powerful reasons to suspect that space has a grainy and atomic structure—but on a scale a trillion trillion times smaller than atoms. This is key unfinished business for twenty-first-century science. According to the most favored theory the fundamental entities are not points but tiny loops, or strings, and the various subnuclear particles are different modes of vibration—different harmonics—of these strings. The particles that physicists study are "woven" from space itself. Moreover, these strings are vibrating not in our ordinary space (with three spatial dimensions, plus time) but in a space of ten or eleven dimensions.

We are three-dimensional beings: we can go left or right, forward or backward, up or down, and that is all. So how are the extra dimensions, if they exist, concealed from us? This may be because they are wrapped up tightly. A long hose-pipe may look like a line (with just one dimension) when viewed from a distance, but from closer up we realize that it is a long cylinder (a two-dimensional surface) rolled up tightly; from still closer, we realize that this cylinder is made from material that isn't infinitely thin, but extends in a third dimension. By analogy, every apparent point in our three-dimensional space, if hugely magnified, may actually have some complex structure: a tightly wound origami in six extra dimensions. Some of the extra dimensions may be more loosely wrapped, so that their effects show up on a microscopic scale in laboratory experiments: indeed there are optimists who believe that the new Large Hadron Collider in Geneva could reveal clues to these extra dimensions.

The microstructure of space manifests itself on scales far smaller than any we can directly probe. Likewise, at the other extreme, our cosmological theories offer intimations that the universe is vastly—perhaps even infinitely—more extensive than the patch we can observe with our telescopes. The domain that astronomers call "the universe"— the space, extending more than 10 billion light-years around us and containing billions of galaxies, each with billions of stars, billions of planets (and maybe billions of biospheres)—could be an infinitesimal part of the totality. There is a definite horizon to direct observations: a spherical shell around us, such that no light from beyond it has had time to reach us since the big bang. But there is nothing physical about this horizon, any more than there is anything special about the horizon here on the Earth. If you were in the middle of an ocean, it's not likely that, just beyond your horizon, the water actually ends. And there are reasons to suspect that our universe—the aftermath of our big bang—extends hugely further than we can see. And that is not all.

Our big bang may not be the only one. Some have speculated that other universes could exist alongside ours. Imagine ants crawling around on a large sheet of paper (their two-dimensional universe). They would be unaware of a similar sheet that's parallel to it. Likewise, there could be another entire universe (with three-dimensional space, like ours) less than a millimeter away from us, but we'd be oblivious to it if that millimeter were measured in a fourth spatial dimension, while we are imprisoned in just three.

It was perhaps self-indulgent to start this chapter with the most remote and speculative topics. But the bedrock nature of space and time, and the structure of our entire universe, are surely among science's great open frontiers. They exemplify intellectual domains where we're still groping for the truth—where, in the fashion of ancient cartographers, we must still inscribe "here be dragons." A unified theory, if achieved, would complete a unification program that started with Newton, who identified the universal force of gravity, and continued through Faraday and Maxwell, who showed that electric and magnetic forces were intimately linked, and their successors. It might even realize the Pythagorean vision of reducing all nature's complexities to geometry. Until we have such a theory we won't understand one of the deepest mysteries that astronomy has revealed—that there is dark energy latent even in empty space, which pushes galaxies apart at an accelerating rate. And our successors will need to address questions that we can't yet formulate: Donald Rumsfeld's famous "unknown unknowns" (what a pity, incidentally, that Rumsfeld didn't stick to philosophy!).

Einstein himself worked on an abortive unified theory till his dying day; in retrospect it is clear that his efforts were premature—too little was then known about the forces and particles that govern the subatomic world. Cynics have said that he might as well have gone fishing from 1920 onward, but there's something rather noble about the way he persevered, reaching beyond his grasp. (Likewise,

Francis Crick, the driving intellect behind molecular biology, shifted, when he reached sixty, to the "Everest" problems of consciousness and the brain even though he knew he'd never get near the summit.)

The cumulative advance of science requires new technology and new instruments—in symbiosis, of course, with theory and insight. The Large Hadron Collider at CERN in Geneva is the world's biggest and most elaborate scientific instrument. Its completion in 2009 generated enthusiastic razzmatazz and wide public interest; but at the same time questions were understandably raised about why such a large investment was being made in seemingly recondite science. But what is special about this branch of science is that its practitioners in many different countries have chosen to commit much of their resources over a time span of nearly twenty years to construct and operate a single vast instrument in a Europe-led collaboration. Britain's annual contribution amounts to about 2 percent of its overall budget for academic science, which doesn't seem a disproportionate allocation to a field so challenging and fundamental (and in which Britain and the United States have a specially strong record, and can aspire to more than their pro rata share of the discoveries). This global collaboration on a single project to probe some of nature's most fundamental mysteries—and push technology to its limits—is surely something in which our civilization can take pride.

Twenty-first-century challenges

We are witnessing stronger links between two frontiers of science: the very large (the cosmos) and the very small (the quantum). But only a tiny proportion of researchers are cosmologists or particle physicists. There's a third frontier, the very complex—and that's where 99 percent of scientists deploy their efforts. Our everyday world presents intellectual challenges just as daunting as those of the cosmos and the quantum. It may seem incongruous that scientists can make confident statements about galaxies billions of light-years away, while being baffled about issues close at hand that we all care about—diet and common diseases, for instance. But this is because our environment is so immensely complicated. Even the smallest insect, with its layer upon layer of intricate structure, is far more complex than either an atom or a star.

The different sciences are sometimes likened to successive levels of a tall building: physics on the ground floor, then chemistry, then cell biology, and all the way up to psychology—with the economists in the penthouse. There is a corresponding hierarchy of complexity: atoms, molecules, cells, organisms, and so forth. But the analogy fails in a crucial respect. In a building, insecure foundations imperil everything above; but the higher-level sciences dealing with complex systems aren't imperiled by an insecure base. The uncertainties of subatomic physics are irrelevant to biologists and environmentalists. To those who study how water flows—why it goes turbulent, or why waves break—it's irrelevant that water is molecules of hydrogen and oxygen. An albatross returns to its nest after wandering 10,000 miles in the southern oceans—and it does this predictably. But it would be impossible, even in principle, to calculate this behavior "bottom up" by regarding the albatross as an assemblage of atoms.

Everything, however complicated—breaking waves, migrating birds, and tropical forests—is made of atoms and obeys the equations of quantum physics. But even if those equations could be solved, they wouldn't offer the enlightenment that scientists seek. Each science has its own autonomous concepts and laws. Reductionism is true in a sense. But it's seldom true in a *useful* sense. Problems in biology, and in environmental and human sciences, remain unsolved because it's hard to elucidate their complexities, not because we don't understand subatomic physics well enough.

Let us focus now on some specifics. If I were to conjecture where the scientific cutting edge will advance fastest, I'd plump for the interface between biology and engineering. Practitioners of the new science of synthetic biology can construct a genome from small stretches of DNA. And another burgeoning discipline, nanotechnology, aims to build up inorganic structures atom by atom, leading to even more compact devices that will enhance computer processing and memory and could enable nanorobots. Computers are already transformational, especially in fields where we can't do real experiments. In the "virtual world" inside a computer, astronomers can mimic galaxy formation or crash another planet into the Earth to see if that's how our Moon might have formed; meteorologists can simulate the atmosphere, for weather forecasts and to predict long-term climatic trends; brain scientists can simulate how neurons interact. Just as video games get more elaborate as their consoles get more powerful, so, as computer power grows, these virtual experiments become more realistic and useful.

Some things, like the orbits of the planets, can be calculated far into the future. But such cases are actually atypical. In most contexts, there's a fundamental limit to how far ahead we can predict. That's because tiny contingencies— like whether a butterfly flaps its wings—have consequences that grow exponentially. For reasons like this, even the most fine-grained computation cannot normally forecast British weather even a few days ahead. (But—and this is important—this doesn't stymie predictions of long-term climate change, nor weaken our confidence that it will be colder next January than it is in July.) So there are limits to what can ever be learned about the future, however powerful computers become. But what can we conjecture, more broadly, about how science will develop in the rest of this century?

Understanding the brain—the most complicated thing we know about in the universe—is of course a supreme challenge. Scanning techniques are revealing how our brains develop, and how our thoughts and emotions are processed. But already new debates are opening up about personal responsibility and freedom. The US National Academy of Sciences recently gave a special award for a project entitled "Neural Correlates of Admiration and Compassion." This is scary. If scanners can reveal our emotions and obsessions, when we are sincere and when we are bluffing, that's the ultimate invasion of our privacy.

One thing that's changed little for millennia is human nature and human character. Before long, however, new cognition-enhancing drugs, genetics, and cyborg techniques may alter human beings themselves. That's something qualitatively new in recorded history—and disquieting because it could portend more fundamental forms of inequality if these options were open only to a privileged few. And we are living longer. Ongoing research into the genetics of aging may explain why—indeed, a real wild card in population projections is that future generations could achieve a really substantial enhancement in lifespan. This is still speculation—mainstream researchers are cautious about the prospect of improvements that are more than incremental. (And of course whether a longer lifespan is indeed an "improvement" depends on whether it is the years of full activity or those of senile decrepitude that are prolonged.)

But such caution hasn't stopped cryonic enthusiasts, worried that they'll die before this nirvana is reached, from bequeathing their bodies to be frozen, hoping that some future generations will resurrect them, or download their brains into a computer. For my part, I'd rather end my days in an English churchyard than a California refrigerator.

Will computers take over? Back in the 1990s, IBM's Deep Blue beat Kasparov, the world chess champion. A mobile phone, suitably programmed, can now beat a grand master; and a more advanced IBM computer (dubbed "Watson") competed successfully with humans on a TV game show. But robots can't yet recognize and move the pieces on a real chessboard as adeptly as a child can. Later this century, however, their more advanced successors may relate to their surroundings (and to people) as adroitly as we do through our sense organs. Moral questions then arise. We accept an obligation to ensure that other human beings can fulfill their natural potential; and we even feel the same about some animal species. But what is our obligation toward sophisticated robots, our own creations? Should we feel guilty about exploiting them? Should we fret if they are underemployed, frustrated, or bored?

Be that as it may, robots surely have immense potential in arenas that humans can't readily reach—in mines, oil rigs, and suchlike. Health care may be aided by nanorobots voyaging inside our bodies. And where they might really come into their own is way beyond the Earth—in aiding the long-term human aspiration to explore outer space.

Probing beyond the Earth

Newton realized that a projectile would escape Earth's gravity, and go into orbit, if it reached a speed of 18,000 miles per hour. But it wasn't, of course, until the 1950s that rockets achieved such speeds; the first artificial satellite, the Soviet *Sputnik*, was launched in 1957, followed by a succession of further launches from the USSR and the United States. Humans soon followed. In the 1960s *manned* space flight went from cornflakes packet to reality. Neil Armstrong's "small step" on the Moon came only twelve years after *Sputnik*—and only sixty-six years after the Wright brothers' first flight.

Had the pace been sustained there would by now have been a lunar base, even an expedition to Mars. But the Moon race was an end in itself, driven by the urge to beat the Russians; there was no motive to sustain the huge investment and maintain the pace of the 1960s. Only the middle-aged can remember when men walked on the moon. Films like *Apollo 13* and *In the Shadow of the Moon* were for me (and I suspect for many others of similar vintage) an evocative reminder of historic episodes that we followed anxiously at the time. But to the young, the outdated gadgetry and right-stuff values portrayed in these films seem as antiquated as those of a traditional Western.

Post-*Apollo*, hundreds of astronauts have circled the earth in low orbits, but none has gone further. Instead, unmanned space technology has flourished, giving us GPS, global communications, environmental monitoring, and other everyday benefits. And scientific exploration has burgeoned too. Probes to Mars and to the moons of Jupiter and Saturn have beamed back pictures of varied and distinctive worlds. I hope that during this century the entire solar system will be explored by flotillas of miniaturized unmanned craft. One can imagine robotic fabricators, building large structures, or perhaps mining rare materials from asteroids.

But will people venture there too? The need weakens with each advance in robots and miniaturization—that's my view, as a *practical scientist*. But as a *human being*, I'm nonetheless an enthusiast for manned missions, as a long-range adventure for (at least a few) humans. The next humans to walk on the Moon may be Chinese: China has the resources, the *dirigiste* government, and, perhaps, the willingness to undertake an Apollo-style program. Americans have downgraded the priority of manned space flight: their firm plans don't even include a return to the Moon. The main impediment for NASA is that it's constrained by public and political opinion into being too risk-averse. The space shuttle failed twice in 135 launches. Although this represents a level of risk that astronauts or test pilots would willingly accept, the shuttle had been promoted as a safe vehicle for civilians. Each failure caused a national trauma and was followed by a hiatus in the program while costly efforts were made (with very limited effect) to reduce the risk still further.

I don't think future expeditions to the Moon and beyond will be politically and financially viable unless they are cut-price ventures, spearheaded by individuals prepared to accept high risks—perhaps even "one-way tickets." And these may have to be privately funded; no Western governmental agency would expose civilians to such a hazardous venture. It is now US policy to encourage private companies to undertake launches—rendering NASA more like an airport authority and less like an airline. The Falcon 9 rocket, developed by the SpaceX company led by the entrepreneur Elon Musk, has successfully launched a payload into orbit. The involvement in space projects of Elon Musk, Jeff Bezos (founder of Amazon), and others in the high-tech community with credibility and resources is surely a positive step. And Google has offered a prize for whoever can build and launch a robotic lunar lander that can carry out specific tasks on the Moon. This is another stimulus—leveraging far more money than the prize itself offers.

There is a step change in cost and technical challenge

between orbital flights that circle the Earth and expeditions to the Moon and beyond. But it's surely not unrealistic to envisage private sponsorship at the multi-billion-dollar level: this is within reach even of some individuals. A comparison might be Formula One car racing, where leading teams have each had budgets of around $400 million a year.

There may be a parallel here with terrestrial exploration, which was driven by a variety of motives. The explorers who set out from Europe in the fifteenth and sixteenth centuries were bankrolled mainly by monarchs, in the hope of recouping the expenditure in exotic merchandise or by colonizing new territory. Some, for instance Captain Cook's three eighteenth-century expeditions, were publicly funded, at least in part as a scientific enterprise. And for some, generally the most foolhardy of all, the enterprise was primarily a challenge and adventure: the same motive that drives test pilots, mountaineers, and round-the-world sailors. And it will be dangerous. Remember that nowhere in our solar system offers an environment as clement, even, as the Antarctic or the top of Everest. It is foolish to claim, as some do, that emigration into space offers a long-term escape from Earth's problems.

A century or two from now, however, small groups of intrepid adventurers may be living independently from the Earth. Whatever ethical constraints we impose here on the ground, we should surely wish such pioneers good luck in genetically modifying their progeny to adapt to alien environments. This might be the first step toward divergence into a new species: the beginning of the posthuman era. And machines of human intelligence could spread still further. Whether the long-range future lies with organic posthumans or with intelligent machines is a matter for debate. Would it be appropriate to exploit Mars, in a manner akin to that of the pioneers who advanced westward across the United States? Should we send seeds for plants genetically engineered to grow and reproduce there? Or should the Red Planet be preserved as a natural wilderness, like the Antarctic? The answer should depend on what the pristine state of Mars actually is. If there were any life there already—especially if it had different DNA, testifying to quite separate origin from any life on Earth—then there would be widely voiced insistence that Mars should be preserved unpolluted.

Is there life out there already?

And this leads to one of the other great unknowns. Do we really expect to find any living creatures out there already? Firm evidence for even the most primitive bugs or bacteria would be immensely significant. But what really fuels popular imagination is the prospect of advanced life—the aliens familiar from science fiction. (I'm discounting, of course, that aliens in UFOs have already visited us. The claimed manifestations—crop circles and the like—are as banal and unconvincing as the messages from the "other side" routinely reported in the heyday of spiritualism a hundred years ago.) Mars is a frigid desert with a very thin atmosphere. There may be simple life there, or remnants of creatures that lived early in the planet's history; and there could conceivably be life, too, in the ice-covered oceans of Jupiter's moon Europa, but nobody expects a complex biosphere in those locations. Suppose, however, we widen our gaze beyond our solar system.

The Italian monk and scholar Giordano Bruno, burned at the stake in 1600, conjectured that the stars were other suns, each with their retinue of planets. Four hundred years later, science confirms this: our Sun is just one star among billions in the vastness of space far beyond our own solar system. Astronomers have learned (but only since the 1990s) that other stars indeed have planets circling around them, just as the Earth, Mars, and Jupiter circle around our own star, the Sun. These planets have not yet actually been seen. The first few hundred to be discovered were inferred indirectly by measuring the wobble induced in the motion of their parent star by their gravitational pull. The extrasolar planets discovered by this technique are very big, rather like Jupiter and Saturn, the giants of our own solar system. But we'll be especially interested in possible twins of our Earth—planets the same size as ours, orbiting other Sun-like stars, on orbits with temperatures such that water neither boils nor stays frozen. Detecting Earth-like planets, hundreds of times less massive than Jupiter, is a real challenge. They induce motions of merely centimeters per second in their parent star—too small for current techniques to measure.

There's another way to search for such planets: we can look for their shadow. A star would dim slightly when a planet was in transit in front of it; an orbiting planet would cause regularly repeating dimming, occurring once per orbit. NASA's *Kepler* spacecraft, launched in March 2009, has been designed to detect this phenomenon. It carries a modest-sized telescope that points steadily at the Cygnus and Lyrae constellations. It monitors the brightness of about 150,000 stars in its field of view, and repeats this every half hour. It is sensitive enough to detect a dimming by just 1 part in 10,000 (which is what would be expected from the transit of a planet whose diameter was 100 times smaller than that of the star). *Kepler* can thereby reveal planets no bigger than the Earth and tell us how commonly they occur. In February 2011, tentative detections of planets around a thousand different stars were released (and one star was discovered to have no fewer than six planets). The planets so far discovered are closer to their parent star than the Earth is to the Sun; it will take another two years to identify planets that are in slower Earth-like orbit, but there is every expectation that these, too, will be numerous.

But we'd really like to see these planets directly—not just

their shadow—and that's hard. To realize just how hard, suppose an alien astronomer with a powerful telescope was viewing the Earth from, say, 30 light-years away—the distance of a nearby star. Our planet would seem, in Carl Sagan's phrase, a "pale blue dot," very close to a star (our Sun) that outshines it by many billions: a firefly next to a searchlight. But if the hypothetical aliens could detect the Earth, they could learn quite a bit about it. The shade of blue would be slightly different depending on whether the Pacific Ocean or the Eurasian land mass were facing them. They could infer the length of the day, the seasons, that there are oceans, the gross topography, and the climate. By analyzing the faint light, they could infer that Earth has a biosphere. Within twenty years, huge telescopes, in space or on the ground, will allow us to draw such inferences about planets the same size as our Earth, orbiting other Sun-like stars.

In his 1584 book *On the Infinite Universe and Worlds*, Bruno's speculations went further: on some of those planets, he conjectured, there might be creatures "as magnificent as those upon our human earth." On this issue, we've little more evidence than Bruno had. Could some of these extrasolar planets harbor life-forms far more interesting and exotic than anything we might find on Mars? Could they even be inhabited by beings that we could recognize as intelligent? Our cosmos would then seem far more interesting: we would look at a distant star with renewed interest if we knew it was another sun, shining on a world as intricate and complex as our own.

We still know too little to say whether alien life is likely or unlikely; indeed the origin of life on Earth is a key unsolved problem. As often in science, lack of evidence leads to polarized opinions in the community, but I think utter open-mindedness is the only rational stance while we know so little about how life might start and what evolutionary paths it might take.

Even the most firmly Earth-bound scientist would accept that one of the great challenges is to understand how life gets started. There's an enormous variety of life on Earth—from slime mold to monkeys (and, of course, humans as well). Life seems to have been present from the very earliest times and survives in the most inhospitable corners of our planet—inside arid desert rocks, deep underground, and in the highest reaches of the atmosphere. We know that all these diverse species, many millions of them, share a genetic code based on DNA. But the basic transition from nonliving to living—the origin of the very first life—is almost as mysterious today as it was in Darwin's time. What led from amino acids to the first replicating systems, and to the intricate protein chemistry of monocellular life? Laboratory experiments that try to simulate the soup of chemicals on the young Earth may offer clues; so might computer simulations. Darwin envisaged a "warm little pond." We are now more aware of the immense variety of niches that life can occupy. The ecosystems near hot sulphurous outwellings in the deep oceans tell us that not even sunlight is essential. So life's beginnings may have occurred in a torrid volcano, deep underground, or even in the rich chemical mix of a dusty interstellar cloud.

Within our solar system, Earth is the Goldilocks planet—not too hot and not too cold. Were it too hot, even the most tenacious life would fry; if too cold then the processes that created life would have happened far too slowly. But the correct temperature is not the only important thing. Everywhere you find life on Earth you find water: not necessarily oxygen, nor always sunlight—but always water. A source of energy and water seem to be the bare necessities for life. Analysis of interstellar space has shown that water is abundant throughout the universe and that starlight is also in great supply. It seems that the basic ingredients are out there, but is there life? The origin of life on Earth might have involved a fluke so rare that it happened only once in the entire galaxy—like shuffling a whole pack of cards into a perfect order. On the other hand, it might turn out that the process was almost inevitable given the right environment. So perhaps the cosmos teems with life.

Incidentally, if any signs of life were found elsewhere in our solar system—and (an important proviso) if we could be sure that it was based on a different kind of DNA, implying that it had a separate origin from terrestrial life—then we could immediately conclude that life was widespread in the universe. Something that had happened twice around a single star must have happened on millions of planets elsewhere in the galaxy.

Or even intelligent life?

Even if simple life is common, it is of course a separate question whether it's likely to evolve into anything that we might recognize as intelligent or complex. Indeed, evolutionists don't agree on how differently our own biosphere could have developed if contingencies like ice ages and meteorite impacts had happened differently. If, for instance, the dinosaurs hadn't been wiped out, the chain of mammalian evolution that led to humans may have been foreclosed and it's not clear whether another species would have taken our role.

Complex biospheres like the Earth's could be rare because of some bottleneck, some key stage in evolution, that is hard to transit. Perhaps it is the transition to multicellular life. (The fact that simple life on Earth seems to have emerged quite quickly, whereas even the most basic multicellular organisms took nearly 3 billion years, suggests that there may be severe barriers to the emergence of any complex life.) Or the bottleneck could come later. Perhaps, more ominously, there could be a bottleneck at our own present evolutionary stage—the stage when intelligent life starts to develop technology. If so, the future development of life on (and perhaps beyond) the Earth depends on whether humans survive this critical phase. This requires avoidance of a cataclysm that wipes us out—unless, before this happens, some humans or advanced artifacts have spread beyond our home planet.

Maybe the search for life shouldn't restrict attention to planets with biospheres like that of the Earth. Science-fiction writers have other ideas: balloon-like creatures floating in the dense atmospheres of Jupiter-like planets, swarms of intelligent insects, nanoscale robots, and more. (And it's often better to read first-rate science fiction than second-rate science—it's far more stimulating, and perhaps no more likely to be wrong.) Indeed it is surprising that depictions of aliens show limited variety—they are predominantly envisaged as mammalian bipeds. The aliens may not be organic at all. The most durable form of life may be machines whose creators have long ago been usurped or become extinct. An intelligent race of aliens could have manufactured self-reproducing machines that spread through the cosmos while their creators stayed at home. The machines could have intelligence—even superhuman intelligence—but they would not necessarily have conscious feelings. And they could be much smaller than us. They may be nanorobots—immensely complex, but almost too small to be seen. Indeed, perhaps they are here already!

The detection of extraterrestrial intelligence would be

an immense culture shock for humanity—it would mean that we were part of a Galactic club and that it would be worth searching for further examples of alien life by all astronomical techniques. On the other hand, it would be a blow to humanity's cosmic self-esteem. However, were our biosphere unique it would disappoint the searchers, but, in compensation, we could be less cosmically modest: our Earth, tiny though it is, would be uniquely important in the galaxy.

We may learn this century whether biological evolution is unique to the pale blue dot in the cosmos that is our home, or whether Darwin's writ runs through a wider universe that teems with life—even with intelligence. But even in the latter case, such intelligence could be unimaginably different from our own. Some "brains" may package reality in a fashion that we can't conceive and may have a quite different perception of reality. Others could be uncommunicative: living contemplative lives, perhaps deep under some planetary ocean, doing nothing to reveal their presence and having no motive for interstellar travel. Still other brains could actually be assemblages of superintelligent social insects. There may be a lot more out there than we could ever detect. Absence of evidence wouldn't be evidence of absence.

In his *Philosophical Investigations*, Ludwig Wittgenstein famously wrote, "If a lion could speak, we couldn't understand him." So would the culture gap with aliens be unbridgeable? Not necessarily. They may come from planet Zog and have seven tentacles but they'd be made of the same kind of atoms as us. If they had developed advanced technology, they would share with us an understanding of physics, math, and astronomy. They'd gaze out, if they had eyes, at the same cosmos—they'd trace their origins back to the same big bang. But they might find string theory a doddle—and understand things that are beyond our grasp.

Could humans eventually understand everything?

This thought takes us back to the question raised at the beginning of this chapter: Are there intrinsic limits to our understanding, or to our technical capability? Could some branches of science come to a halt because we bump up against the inherent limits of our brainpower, rather than because the subject is exhausted? Humans are more than just another primate species. We are special: our self-awareness and language were a qualitative leap, allowing cultural evolution, and the cumulative diversified expertise that led to science and technology. But some aspects of reality—a unified theory of physics, or a full understanding of consciousness—might elude us simply because they're beyond human brains, just as surely as Einstein's ideas would baffle a chimpanzee. Perhaps complex aggregates of atoms, whether brains or machines, can never understand everything about themselves.

Simulations, using ever more powerful computers, will extend scientists' capacity to understand processes that can be neither studied in our laboratories nor observed directly. Future discoveries may be made by brute force rather than by insight. Computers with human-level capabilities will accelerate science, even though they won't think like we do. Deep Blue beat Kasparov by exploiting its higher processing speed to explore millions of alternative series of moves and responses before deciding an optimum move; likewise, machines will make scientific discoveries that have eluded unaided human brains. For example, some substances are perfect conductors of electricity when cooled to very low temperatures (superconductors). There is a continuing quest to find the recipe for a superconductor that works at ordinary room temperatures (that is nearly 300 degrees above absolute zero; the highest superconducting temperature achieved so far is 120 degrees). This quest involves a lot of trial and error, because nobody understands exactly what makes the electrical resistance disappear more readily in some materials than in others.

Suppose that a machine came up with such a recipe. It might have succeeded in the same way that Deep Blue trounced Kasparov, by testing millions of possibilities rather than by having a theory or strategy. But it would have achieved something that would get a scientist a Nobel Prize. Moreover, its discovery would herald a technical breakthrough that could, among other things, lead to still more powerful computers—an example of the runaway acceleration in technology, worrying to some futurists, that could be unstoppable when computers can augment or even supplant human brains.

Toward the far future

One final question: Are there special perspectives that astronomers can offer to science and philosophy? I think there are. Astronomers are disclosing insights that New Agers would welcome and be attuned to. Not only do we share a common origin, and many genes, with the entire web of life on Earth, but we are linked to the cosmos. All living things depend on the stars: they are energized by the heat and light from the Sun; they are made of atoms that were forged from pristine hydrogen, billions of years ago, in faraway stars.

More significantly, astronomers can offer an awareness not only of the immensity of space but also of the immense time spans that lie ahead. The stupendous time spans of the evolutionary past are now part of common culture (apart from in creationist circles). Our present biosphere is the outcome of about 4 billion years of evolution. But most people still somehow think we humans are necessarily the culmination of the evolutionary tree. That hardly seems credible to an astronomer, aware of huge time horizons extending into the future as well as into the past. Our Sun formed 4.5 billion years ago, but it's got 6 billion more before the fuel runs out. And the expanding universe will continue—perhaps forever—becoming (according to the best current long-range forecast) ever colder, ever emptier. As Woody Allen said, "Eternity is very long, especially towards the end." So, even if life were now unique to Earth, there would be scope for posthuman evolution—whether organic or silicon-based—on the Earth or far beyond.

It won't be humans who witness the Sun's demise; it will be entities as different from us as we are from a bug. We can't conceive what powers they might have. But there are some things they couldn't do, like travel back in time. So they can never tell us what (if anything) still perplexes them.

4

A RUNAWAY WORLD

In 2002 three mathematicians at Kanpur in India, Manindra Agrawal and his two students Neeraj Kayal and Nitin Saxena, made a breakthrough that was important for codes and code-breaking. They posted their results on the web. Within just a day, 20,000 people had downloaded the work, which became the topic of hastily convened discussions in research centers around the world.

This episode—offering instant global recognition to two Indian students—contrasts starkly with the struggles of a young Indian a hundred years ago. Srinivasa Ramanujan, a clerk in Mumbai, mailed long screeds of mathematical formulas to G. H. Hardy, a professor at Trinity College, Cambridge. Fortunately, Hardy had the percipience to recognize that Ramanujan was not the typical green-ink scribbler who finds numerical patterns in the Bible or the pyramids, but that his writings betrayed preternatural insight. Hardy arranged for him to come to Cambridge and did all he could to foster his genius—sadly, however, poor health led Ramanujan to an early death.

This anecdote introduces my final topics of discussion: how to optimize science and innovation in the era of global mobility and networking, and how the benefits of global-ization can be shared, to mutual benefit, by developing nations as well as the developed world (and I'll venture some sociological digressions along the way).

The Open University (which hosted my final Reith Lecture) was a visionary creation of the 1960s. It was inspired by the great "social inventor" Michael Young, and stands as the most durable legacy of two Labour politicians, Harold Wilson and Jenny Lee. But its pioneers couldn't have conceived, in the era of black-and-white TV, how fundamentally IT would change the way we learn—how it could globalize students' horizons, level the playing field between those in major centers and those in relative isola-tion, and give outstanding teachers a worldwide reach. And the capability to access huge data sets has transformed not only science but also finance and all international business.

Migration and clustering

But enterprising individuals aren't content to be linked merely in cyberspace. They still tend to swarm together most conspicuously in high-tech hotbeds like Silicon Valley and around the world's leading research universities. The United States benefits hugely by draining highly skilled migrants from the rest of the world. And its allure has been enhanced by the rhetoric and the initiatives of the Obama administration, which has boosted America's already world-leading scientific community—in morale and in substance.

Britain is of course a far smaller player, but it punches above its weight scientifically: it's number two to the United States in most indices, and may be number one in brainpower for the buck, when allowance is made for its lower resources and salaries. But its success in attracting and retaining mobile talent is now at risk.

Indeed, to retain international competitiveness even the United States must raise its game, to confront the strengthening challenge from the Far East. India produces, each year, tens of thousands of highly motivated graduates. That's why 100 global companies have set up R&D centers in India and why Intel chips are being designed there: India could become a scientific superpower by 2020. And, of course, China could surge further ahead still.

There's now an international market for the best students as well. They are academic assets, and a long-term investment in international relations. After they graduate they'll feed into all walks of life, networked worldwide, ready to seize the best ideas from anywhere and run with them.

Migrants can now, unlike a century ago, retain contact with their homeland: communications are always open; travel is far easier and cheaper. In fact, there's a growing two-way traffic. Some now use the term "brain circulation," rather than "brain drain" to describe what's happening in China, India, Taiwan, and Ireland. But there's no such consolation for the least-developed countries: they face daunting challenges in retaining their all-too-few highly trained people, and even more in attracting them back. Those of us in the developed world should surely be uneasy about this and feel the obligation to redress this loss.

Africa's predicament is the worst. Around half of its health workers want to leave and their departure can be ill-afforded; it's doubly tragic if, after moving to a developed country, they find they're not accredited, and doctors become cab drivers. It's just as bad in agricultural science, engineering, and all the other specialities that African countries require if they are to develop their potential. The poorest countries need to engage their diaspora communities, encouraging those with expertise to at least make regular visits back home. But wealthier nations should take some responsibility too. A cost-effective form of aid would be to establish, in Africa and elsewhere, centers of excellence—with strong international links—where ambitious scientists could work in less dispiriting conditions, perhaps via linkages with foreign experts. They could then fulfill their potential without emigrating, and strengthen tertiary education in their home country.

There are encouraging initiatives. My former colleague Neil Turok spearheaded the African Institute of Mathematical Sciences in Capetown—a pioneering institution that offers postgraduate courses to students from all over Africa (at far lower cost, incidentally, than could be achieved in Europe or the United States). This is a model now being replicated in Senegal, Ghana, and elsewhere. It would seem equitable, too, that, for each skilled person drained to the developed world, the receiving country should feed back sufficient resources to train two more.

Of course the real tectonic shift in the world's science stems from burgeoning growth in the Far East, in China above all. Since 1999, China's R&D spend has risen by 20 percent per year—up to a level that's now second only to the United States'. In the Beijing Olympics, China hugely improved on its record in earlier Games and topped the medal table. This success came from targeting medal-rich sports like gymnastics, shooting, and judo. Likewise, China's technocratic leadership has astutely targeted its priority scientific investment on growth areas. Look, for instance, toward the city of Shenzhen. There, a 500-strong research team is hard at work, on the front line of genetic research. They were established only eleven years ago and now have more gene-sequencing capacity than anywhere in the world—enough to sequence 10,000 human genomes in a year. And China strives to lead, too, in solar power—a quite different but equally burgeoning field. In China's latest five-year plan, the strategic emerging industries—among which clean energy technologies are prominent—are projected to grow at more than 20 percent per annum.

Universities and high-tech hubs

Education is prioritized in China—and in Taiwan, South Korea, and other countries of the Far East—with a focus on fulfilling the aspirations of their fast-developing economies. At the university level, the United Kingdom is still well placed. British universities have problems but are in far better shape than those in most comparable countries. In particular, they compare well with those in the larger countries of mainland Europe; they have more diversity of funding, more diversity of mission, and more autonomy over admissions of students and of governance. One should be properly cynical about the spurious precision of the various league tables, but it counts for something that Britain is the only country outside the United States with several universities ranked in the premier league.

Indeed, what we see in mainland Europe reinforces the virtues of the research university model that prevails preeminently in the United States. Britain is fortunate to have evolved a similar model and should cherish it. Research universities benefit the economy partly through direct knowledge transfer from university labs to industry, but they offer an indirect benefit that's harder to quantify but could be even more important. They're a source of independent expertise, of people who are plugged in to new ideas emanating from anywhere in the world. Such universities, harboring the best research teams, can be more effective than freestanding labs in engendering direct knowledge transfer because they are interdisciplinary.

Great institutions like Harvard and Berkeley are major national assets through the worldwide pull they exert on talent, the collective expertise of their faculty, and the consequent quality of the graduates they feed into all walks of life, and in the high-tech clusters that spring up around them. In such environments, individuals feel linked into a supportive network, so that if their company folds they can readily find another niche without leaving the neighborhood.

Although start-up funding remains tighter, the entrepreneurial culture associated with the US West Coast is developing in Britain—though it is regrettable, indeed shameful, that our efforts lag those of a much smaller country such as Israel. A dynamic high-tech community has grown up around Cambridge University that offers, in the words of the *Financial Times*, a "low risk place to do high risk things." In such clusters success breeds success—and, just as important, failure is accepted as a step toward later success. Talent attracts talent (and big companies too).

Britain would benefit if the universities in the rest of Europe were to strengthen, and if there were greater mobility and opportunity between the different European centers. Europe collectively could then offer a stronger counterattraction to the United States as a destination of choice for mobile talent than Britain alone ever could. (Indeed there has been a beneficial change since the 1970s. At that time, young British scientists tended to meet their European counterparts because all went on fellowships to the United States. Now there is more interchange between European countries—and, of course, with China, India, Japan, Korea, and Singapore.)

In the so-called big sciences, which require international-scale facilities, there's long been well-managed European collaboration, and European consortia have achieved real excellence. CERN in Geneva, home of the Large Hadron Collider, is destined to be the world's leading laboratory in particle physics for at least the next fifteen years. The European Southern Observatory has an array of four 8-meter telescopes on a superb site in Chile, and is now planning a giant instrument with a mosaic mirror 39 meters in diameter. Europe has never had a space program to match that of America but could gain an ascendancy even in that arena if it focused on science, miniaturization, and robotics, leaving NASA to squander its far larger budget on a manned program that is neither useful nor inspiring. These grand flagship projects aren't, of course, typical of research. But they're good portents: they show that Europe can fully match the United States by optimally developing a collaborative research community.

The countries of the Far East still have some way to go before they can match the higher-education system of the United States or even of Britain, although their leading research universities are advancing at a spectacular pace. But these nations are already ahead of the West in school-level education. Attainment levels of British and American pupils are poor by international standards—a depressing augury for the future. This weakness was highlighted in an influential 2005 report for the National Academy of Sciences entitled *Rising Above the Gathering Storm*. There are not enough good science teachers to ensure that every pupil gets exposed to one. In consequence, at least half of our young people are deprived of the level of education that prepares them for the more demanding university courses. Despite many initiatives, and some positive trends, substantial improvements will be slow. In the short run there are three things we can do: ensure that conditions are good enough to retain excellent teachers; encourage mature professionals to move into teaching from a career in research, industry, or the armed forces; and make better use of the web and distance learning.

Sustaining the enthusiasm of the young

The young have a natural interest in science, whether focused on space, dinosaurs, or tadpoles. The challenge for educators—and one where most Western nations are failing—is to sustain this interest beyond the primary-school stage. Apart from the need for good teachers another thing that's crucial is hands-on involvement—showing, not just telling. And the sophistication of modern technology is, ironically, an impediment to engaging young people with it.

Newton, when young, made model windmills and clocks—the high-tech artifacts of his time. Darwin collected fossils and beetles. Einstein was fascinated by the electric motors and dynamos in his father's factory. Fifty years ago, inquisitive children could take apart a radio set or motorbike, figure out how it worked, and even put it together again. But it's different today. The gadgets that *now* pervade young people's lives, mobile phones and suchlike, are baffling black boxes—pure magic to most people. Even if you take them apart you'll find few clues to their arcane miniaturized mechanisms. And you certainly can't put them back together again. There's now, for the first time, a huge gulf between the artifacts of our everyday life and what even a single expert, let alone the average child, can comprehend. And the increasing concentration of populations into cities means that fewer people have direct contact with the natural world. These trends present obstacles to even the best teachers—but, of course, technology has provided compensating benefits, especially through ubiquitous computers and the web.

Science education isn't just for those who will use it professionally. Indeed, everyone should have a feel for science—for cultural reasons, and because they otherwise can't, as citizens, participate in discussing how science is used (see Chapter 1). But if the United States and Europe are to sustain economic competitiveness, enough of their young people need to attain professional-level expertise, as millions now do each year in the Far East. A few, the nerdish element, will take this route, come what may—I'm a nerd myself. But a country can't survive on just this minority of "obsessives." At a time when Britain, in particular, needs to reduce its dependence on the financial sector and rebalance toward high-tech manufacturing and services, the sciences must attract a share of those who are ambitious and have flexible talent—those who have a choice of career paths and who are mindful that the City of London still offers Himalayan salaries, if no longer such high esteem. It's crucial that the brightest young people—savvy about trends and anxiously choosing a career—should perceive cutting-edge science and engineering as an attractive option.

More role models would help. At the start of the last decade, BBC TV ran a series of programs to identify "100 Greatest Britons." The advocates of Darwin and Newton slugged it out among the final six. But there was a third contender—Brunel—and he did better than either, for two reasons. First, his advocate was Jeremy Clarkson, a high-profile TV presenter. Second, there were rumors that students from Brunel University (in London) had voted early and often. Nonetheless, this was an all-too-rare instance of a great engineer being publicly acclaimed. (The other finalists were, by the way, Churchill, Shakespeare, and Princess Diana.)

But Brunel is long dead; scientists and engineers still living deserve wider acclaim. (A few, of course, do become at least B-list celebrities, but generally as TV personalities and not via their prime achievements.) To attract the next generation, prominent role models would be helpful—preferably not all male, gray, and stale. And we must proclaim science both as an intellectual challenge and as a prerequisite for meeting humanitarian imperatives—health, education, and clean energy for the developing world.

A digression into sociology

The title I chose for this chapter, "A Runaway World," was used by the sociologist Anthony Giddens for his 1999 Reith Lectures on globalization—and also, indeed, back in 1967, by the anthropologist Edmund Leach for his contribution to the series.

Leach adopted a provocatively radical stance that resonated with the youth culture of the time. The average BBC listener might have bristled at his view that "only those who hold the past in complete contempt are ever likely to see . . . the new Jerusalem." Indeed Leach wouldn't have carried many scientists with him on this point—science is a cumulative achievement in which we build on the past.

But Leach also said, "It is the young adults, not the old ones, who possess the kind of knowledge that young people need to share." Had he lived to see the ascendancy of Microsoft and Apple, of Google and Facebook—to witness how a cohort of young scientific entrepreneurs have changed the world—he'd surely have felt vindicated: the Internet is a prime exemplar of benign globalization. Let us digress briefly to consider how science looks from the perspective of sociology: How do those of us in the zoo appear when we are the subjects of fieldwork by anthropologists?

The scientific tribe is a fascinating topic for study. Science is an intensely interactive activity. Certainly, what I've enjoyed most in my own career is the involvement in ongoing debates that have gradually clarified perplexities and expanded the area of consensual understanding. It is important, as well as enlightening, to appreciate how pervasive the social and political factors that drive and direct science are. The way scientists work, what problems attract their interest, what styles of explanation are culturally appealing—and (more mundanely) what fields attract funding—plainly depend on a range of political, sociological, and economic factors. These change over time, and create different pressures and demands in different countries. Some projects, especially big international ones like the space program, are a by-product of activities driven by other imperatives. However—and this is crucial—the outcome of scientists' efforts is objective: it can be evaluated by criteria that don't depend on how these ideas were motivated and arrived at. How science is applied, however, *is* a culture-dependent matter.

The physicist Steven Weinberg, in his book *Dreams of a Final Theory*, has given an apt metaphor for scientific breakthroughs: "A party of mountain climbers may argue over the best path to the peak, and these arguments may be conditioned by the history and social structure of the expedition, but in the end either they find a good path to the summit or they do not, and when they get there they know it." By analogy, it is fascinating to study how social and economic factors molded the development of music— for instance, the shift from liturgical to operatic genres, the increase in the scale of orchestral compositions after the transition from private patronage to public concerts, and so on. But such studies, though worthwhile in their own right, are peripheral to the essence of the music itself.

When children (or cartoonists) want to depict an archetypal scientist, they frequently draw a wild-looking male figure, resembling the familiar image of Einstein. I recall a cartoon showing Einstein standing in front of a blackboard.

On it is written $E = ma^2$—crossed out; then $E = mb^2$—also crossed out; and then $E = mc^2$—eureka! Of course, that's not how it happened. Einstein's *annus mirabilis* was 1905. As well as discovering the equation that everyone knows, he wrote three other papers that year, any one of which would have established his reputation. He was then twenty-six years old, working from nine to five in the Swiss patent office as a "technical expert, 3rd class." Pictures from that time show him as a rather dapper young man. But it's the *old* Einstein who's become the iconic figure—the benign and unkempt sage of posters, T-shirts, and cartoons.

In impact on our perception of the physical world Einstein is matched only by Isaac Newton (and, in the biological sciences, of course, by Darwin). But in charisma, there's no contest. Newton was an unappealing character: solitary and reclusive when young; vain and vindictive in his later years. Einstein would have been better company. It's fortunate that the most famous twentieth-century scientist projected an engaging image: a genial figure, ready with an aphorism, and idealistically engaged with the world's problems. But the popular perception of Einstein has a downside. First, it typecasts scientists as eccentric, elderly men—I'm afraid some of us are, but, luckily, most aren't. Second, it unduly exalts armchair theory. Science wouldn't have gotten far by pure thought alone: we're no wiser than Aristotle was. It's developed in symbiosis with advancing technology, from telescopes to computers.

Indeed, Einstein's singular fame as a *pure* scientist over-shadows creativity in applied science. Those who've given us today's amazing technologies deserve equal acclaim. It's good that nearly everyone in Britain has heard of Brunel, and nearly all Americans have heard of Edison, but sad that fewer can name any present-day engineer. (Indeed, the engineering profession is even worse at PR than academic scientists are, otherwise their leading practitioners should surely have the same glamorous profile as our most celebrated architects.)

Doing science doesn't involve any mode of thinking distinct from problem solving by engineers or detectives. The greatest scientists don't fall into a single mold. Some are brilliant. Newton's mental powers seem to have been really off-scale. His concentration was as exceptional as his intellect: when asked how he cracked such deep problems, he said, "by thinking on them continually." In contrast, Darwin was modest in his self-assessment: he wrote in his autobiography: "I have a fair share of invention, and of common sense or judgement, such as every fairly successful lawyer or doctor must have, but not, I believe, in any higher degree."

And when his friend Asa Gray, the Harvard botanist, asked him about religion, Darwin diffidently responded, "The whole subject is too profound for the human intellect. A dog might as well speculate on the mind of Newton. Let each man hope and believe as he can" (a glaringly different stance from some of his present-day disciples!). Of course, we should all oppose, as Darwin did, views manifestly in conflict with the evidence, such as creationism. But we shouldn't set up this debate as religion versus science, but strive instead for peaceful coexistence with mainstream religions, which number many excellent scientists among their adherents. My personal view—a boring one, for those who wish to promote constructive dialogue between science and religion—is that, if we learn anything from the pursuit of science, it is that even something as basic as an atom is quite hard to understand. This should induce skepticism about any dogma, or any claim to have achieved more than a very incomplete and metaphorical insight into any profound aspect of our existence. So I have no religious beliefs; however, I respect the customs and rituals of the Anglican church in which I was raised—just as many brought up in the Jewish tradition see value in sustaining these traditions even though they would call themselves atheists.

If teachers tell young people that they can't have God *and* Darwinism, many will choose to stick with their religion, and be lost to science. Moreover, those of us who regard fundamentalism as a real danger need all the allies we can muster against it. The mainstream churches should be welcomed as being on our side in any such confrontation. (For expressing such views, I am described on Richard Dawkins's website as a "compliant Quisling"!)

Science involves hard slog, but that's not enough: insights—eureka moments—are crucial, too. There are parallels with creativity in the arts; but there are differences, too. Any artist's work is individual and distinctive but it generally doesn't last; contrariwise, even the journeyman scientist adds a few durable bricks to the corpus of public knowledge. But our scientific contributions lose their identity: if A didn't discover something, in general B soon would—indeed, there are many instances of near-simultaneous discovery. Not so, of course, in the creative arts. As Peter Medawar remarked, when Wagner diverted his energies for ten years, in the middle of the Ring cycle, to compose *Die Meistersinger* and *Tristan*, he wasn't worried that someone would scoop him on *Götterdämmerung*.

Even Einstein exemplifies this contrast. He made a greater and more distinctive imprint on twentieth-century science than any other individual; but had he never existed, all his insights would by now have been revealed, though gradually, and probably by several people rather than by a single great mind. Einstein's fame extends far beyond science; he was one of the few in the field who really did achieve public celebrity and is as much an icon of creative genius as Beethoven. His impact on general culture, however, has been ambivalent. It's a pity, in retrospect, that he called his theory "relativity." Its essence is that the local laws are *just the same* in different frames of reference. "Theory of invariance" might have been a more apt choice, and would have staunched the misleading analogies with relativism in human contexts. But in terms of cultural fallout he's fared no worse than others. Heisenberg's uncertainty principle—a mathematically precise concept, the keystone of quantum mechanics—has been hijacked by adherents of oriental mysticism. And Darwin has likewise suffered tendentious distortions, especially in applications to human psychology.

Optimizing scientific creativity

I'm fortunate to know many of today's leading scientists—those who have achieved Nobel-level breakthroughs. They are individualists but have some things in common. They all staked their careers on a specific line of research—and they chose well. The path they took was unpredictable, and often the payoff was long in coming. The winners of the 2010 Nobel Physics Prize, Andrei Geim and Konstantin Novoselov, are exemplars. These two Russian scientists, who have been based for the last decade at Manchester University in England, made the unexpected discovery that carbon atoms could form a lattice just one atom thick—a new material, graphene, with extraordinary strength and many potential uses. Their work didn't need major equipment; the clinching experiment involved a piece of adhesive tape. But these men staked several years of their lives, and their reputation, on their quest. And Manchester offered the security and intellectual freedom they needed.

Universities won't stay internationally competitive unless they can attract and nurture such people. Scientists are of course accountable to their funders, but it's essential that those with a good track record can follow their own judgement rather than be constrained by narrow external targets. And those arguing for such permissiveness aren't being self-indulgent; history shows that it's through free inquiry that research pays the greatest dividends overall.

We can't confidently predict how, when, or whether a specific research project will pay off. The social or economic benefit should not be credited solely to the most immediately relevant project, any more than a win at football is due solely to the team members who actually score the goals. But successes emerge only in a nurturing environment. The major insights in science come from people who have the patience to develop an intimate understanding of a problem, who have the space and the freedom to take professional risks, and who know how to make creative use of the surprises that they encounter when they do so. These are the people whom we must nurture wherever we find them. Confidence and high morale drive creativity, innovation, and risk-taking—whether in science, the arts, or entrepreneurial activity.

In research, second-rate work counts for very little. Researchers have the best expertise for judging what topics hold promise, and the strongest possible motive for choosing an area in which they'll have an impact. The difference in payoff between the very best research and the merely good is, by any measure, thousands of percent. So what is most crucial in giving taxpayers enhanced value for their money isn't the few percent savings that might be made by improving efficiency in the office management sense. It's maximizing the chance of the big breakthroughs by attracting and supporting the right people, and backing the judgment of those with the best credentials.

In lively research groups it's exhilarating when coffee-time conversations toss out new ideas and debate the latest discoveries. The best institutes all foster such an atmosphere, and I'm lucky to work in one of them at Cambridge University. But even in this kind of privileged environment, my colleagues seem ever more preoccupied with grant cuts, proposal writing, job security, and suchlike. Prospects of breakthroughs will plummet if such concerns prey unduly on the minds of even the very best young researchers. Not just in Britain, but throughout the EU, in the United States and elsewhere, bodies that allocate public funds for science and education focus on ever more detailed performance indicators to quantify the output that results. This has the best of intentions—to raise standards, improve accountability, and enhance the chances of beneficial spin-off. But its actual consequences are often the reverse—to impede the best professional practice. Indeed Ben Martin and Puay Tang at Sussex University have identified seven channels of benefit from publicly funded research, of which direct spin-off is just one. They argue—as many have before them—that, taking all seven together, university research offers incontrovertible benefits to the economy and to society. But they note, "A danger that a focus on the more easily measurable exploitation channels . . . may distort policy, to the detriment of longer term benefits."

There are concerns that, unless nongovernmental funding can be greatly expanded, the intellectual atmosphere will become corrosive to open-ended enquiry. Scientific knowledge is collective, public, and international—it is, in principle, accessible to the entire world. But its benefits can only actually be captured by those who are educated and discerning enough—who are plugged in to the research community. That's why it's in the interests of each country to maintain strong and broad expertise; once the tap has been turned off, it can't readily be turned on again.

In his State of the Union Address in January 2011, President Obama urged the need for greater investment in science and technology. He recalled how the Apollo program in the 1960s, a response to Soviet advances in space, had provided a broad impetus to technology and education, and he asserted that his nation faced another Sputnik moment. He offered a neat metaphor: "You can't make an overweight aircraft more flight-worthy by removing an engine." This message is even more vital for countries like my own, which must rebalance its economy away from overdependence on the financial sector, and acknowledge science and innovation as essential engines for long-term prosperity and confronting global challenges.

A British digression

Here is another quote, this time from Britain: "We are supposed to be the clever country. We used to be the commonsense country. Not for much longer if the politicians continue to undervalue the potency of those Francis Bacon called the 'merchants of light', of new knowledge, especially scientific knowledge, which is unarguably the only sure wealth of the future." These words come not from a politician but from a lecture by the writer and journalist Melvyn Bragg, who has himself done more than anyone to bridge the cultures of the arts and the sciences.

For us in Britain, sustaining our scientific standing, with all the economic and social benefits this brings, is crucial. What is needed is a long-term plan to ensure science can share the fruits of the recovery that it will help to generate. We don't know what the twenty-first-century counterparts of the electron, quantum theory, the double helix, and the computer will be, nor where the great innovators of the future will get their formative training and inspiration. But one thing seems clear: we will decline unless we can sustain a competitive edge in discovery and innovation and ensure that some of the key creative ideas of the twenty-first century germinate and—even more—are exploited here.

Our university system stands up to international comparisons better than its schools do, but here, too, there is no room for complacency or conservatism. Some restructuring is needed—indeed, it's overdue. Total enrollment in full-time higher education has risen from less than 10 percent in the 1960s to around 40 percent today. But this welcome expansion hasn't yet led to sufficient diversity. We can learn here from the United States, which is of course home to several thousand institutions of higher education: junior and regional colleges, top-quality liberal arts colleges, huge state universities (many world-class), and the Ivy League private universities. The British system needs to evolve in a similar fashion. It must diversify; its curriculum must also become less specialized. The traditional three- or four-year honors degree course is not appropriate for all present-day students; indeed, a broader curriculum could benefit all students. A wider variety of courses is needed. Students who leave university after just two years should be given credit (saying that they had two years of college) and the chance to return later to the same institution or a different one, rather than being typecast as "wastage."

Britain has no real counterpart to the Ivy League: even Oxford and Cambridge depend predominantly on government funding. Although the Ivy League is widely admired, it retains some features that would, rightly or wrongly, encounter principled resistance even in traditional universities in Britain—for instance, admission procedures that offer an inside track to the offspring of alumni or benefactors. Equally admirable, and in my view a better and more relevant model to emulate, is the system established in California, under the visionary leadership of Clark Kerr. Its three-level structure of colleges embodied (at least throughout the period when it was adequately funded) an enviable combination of excellence, outreach, and flexibility.

The impact of the Internet

Enhanced communication and computer networks are transforming how information is spread and shared. For example, in the old days, astronomical information, even if in principle publicly available, was stored on delicate photographic plates: these were not easily accessible, and tiresome to analyze. Now, such data (and, likewise, large data sets in genetics or particle physics) can be accessed and downloaded anywhere. Experiments, and natural events such as tropical storms or the impact of a comet on Jupiter, can be followed in real time by anyone who is interested.

Moreover, the Internet allows new styles of research. For example, in the Galaxy Zoo project, images of 3 million galaxies can be viewed on the web, and the labor-intensive task of classifying them is being shared by thousands of keen amateur astronomers. And in biology, 40,000 PlayStation 3 enthusiasts are exploring the combinatorial options for protein folding, via the Folding@home website. More surprisingly, wiki-style activity may catch on in mathematics. On the website of my Cambridge colleague Tim Gowers, theorems have been proved via a genuine collective effort, like completing a jigsaw, or the development of open-source software. These are just instances of how scientific progress is being enhanced by the involvement of millions of people worldwide.

(Some pessimists argue that scientific progress will clog up because of information overload. I don't think that's a serious worry. Novel advances bring with them a flood of new data, but at the same time more patterns and regularities are revealed, which cut down the number of disconnected facts worth remembering. There's no need to record the fall of every apple, because, thanks to Newton, we understand how gravity pulls everything—whether apples or spacecraft—toward the Earth.)

As we live longer, in a faster-changing environment, the importance of distance learning will surely grow, worldwide. And students will become dissatisfied with traditional lectures, of the kind offered to passive audiences in universities all over the world. Given, for instance, the choice between viewing the brilliant Harvard course on Justice by Michael Sandel on the Internet or attending mediocre live lectures on the same theme, few would opt for the latter. Indeed, the Internet can offer star lecturers (and even the best classroom teachers) a potentially global reach.

Purely online resources can never fully replace the personal interaction with a tutor or mentor—but their scope and reach will surely expand. Other institutions have already followed MIT's lead in making videos of lecture courses available for viewing online. And there are private innovations too. For instance, the scientifically educated financier Salman Khan has created the Khan Academy: over 2000 ten-minute videos, explaining key concepts in mathematics and other subjects. This is an amazingly cost-effective way to supplement the efforts of teachers and enrich the regular curriculum. Stanford University has gone a step further. A popular course on artificial intelligence given by two eminent professors, Peter Norvig and Sebastian Thrun, is webstreamed, enabling anyone to view the twice-weekly lectures and participate in interactive quizzes. About 190,000 people, worldwide, signed up, and—even more encouragingly—23,000 completed the course.

All these approaches are sure to be quickly emulated—to the benefit, especially, of the millions who never have the opportunity for personal contact with a first-rate teacher or world-class expert.

Conclusion: Sustaining a long-term global vision

There you have it—the perspective of a British scientist, concerned that his country optimizes its competitive advantages, to the benefit not only of itself but also of the wider world as well. But it would be perverse to end on a jingoistic note. Science has always crossed national boundaries. Back in the 1660s, the Royal Society proclaimed its intention to promote "commerce in all parts of the world with the most curious and philosophical persons to be found." Benjamin Franklin urged that Captain Cook's voyage should not be impeded during the war of American independence. Humphrey Davy traveled freely in France during the Napoleonic wars; Soviet scientists retained contact with their Western counterparts throughout the Cold War. And today any leading laboratory, whether it's run by a university or by a multinational company, contains a similarly broad mix of nationalities wherever it is located. Most scientists have peripatetic careers, and, as I emphasized earlier, we need to share and spread expertise throughout the developing world.

Collaborations straddle today's deepest political divides. To quote just one instance, a physics facility (with the acronym SESAME) is being built in Jordan, with support from countries across the Middle East. A project that brings Iran and Israel to the same table on an equal footing could be acclaimed a success on political terms alone. In the coming decades, the world's intellectual and commercial center of gravity will move to Asia, as we see the end of four centuries of European and North American hegemony. But we're not engaged in a zero-sum game. We should welcome an expanded and more networked world, and hope that other countries follow the example of Singapore, South Korea, and, of course, India and China.

And individual scientists act as global citizens. For instance, John Sulston, the Nobel Prize–winner who in the 1990s led the British part of the human genome project, now campaigns to provide affordable drugs for Africa. The great ecologist E. O. Wilson promotes the "Atlas of Life" to document and preserve biodiversity. And, of course, many less-known figures are active in NGOs, government agencies, and global organizations like the World Health Organization and the IPCC.

It's imperative to multiply such examples. We need a change in priorities and perspective—and soon—if the world's people are to benefit from our present knowledge and the further breakthroughs that this century will bring. We need urgently to apply new technologies optimally and to avoid their nightmarish downsides: to stem the risk of environmental degradation; to develop clean energy and sustainable agriculture; and to ensure that we don't in 2050 still have a world where billions live in poverty and the benefits of globalization aren't fairly shared. It's gratifying (though of course unsurprising) that those in the younger generation take a more global perspective and are concerned with long-term environmental issues. It should be the goal of educators to stimulate such attitudes and commitment.

I'll close with a personal perspective. In the flat fenlands of Eastern England, just 15 miles from where I live, is the small town of Ely. It is dominated by a magnificent cathedral, dating from the twelfth century. This immense and glorious building overwhelms us today. But think of its impact 900 years ago. Think of the vast enterprise its construction entailed. Most of its builders had never traveled more than 50 miles—the Fens were their world. Even the most educated knew of essentially nothing beyond Europe. They thought that the world was a few thousand years old and that it might not last another thousand. But despite these constricted horizons in both time and space, despite the deprivation and harshness of their lives, despite their primitive technology and meager resources, they built this cathedral—pushing the boundaries of what was possible. Those who conceived it knew that they wouldn't live to see it finished. Their legacy still elevates our spirits, nearly a millennium later.

What a contrast to so much of our discourse today! Unlike our forebears, we know a great deal about our world—and, indeed, about what lies beyond. Technologies that our ancestors couldn't have conceived now enrich our lives and our understanding. Many phenomena still make us fearful, but the advance of science spares us from irrational dread. We know that we are stewards of a precious pale blue dot in a vast cosmos—a planet with a future measured in billions of years, whose fate depends on humanity's collective actions this century. But all too often the focus is short-term and parochial. We downplay what's happening even now in impoverished faraway countries. And we give too little thought to what kind of world we'll leave for our grandchildren.

In today's runaway world, we can't aspire to leave a monument lasting a thousand years, but it would surely be shameful if we persisted in policies that denied future generations a fair inheritance.

ACKNOWLEDGMENTS

In 2010 the Royal Society celebrated its 350th anniversary. To mark this event the BBC designated 2010 as a "year of science" and commissioned several special programs with scientific themes. It was also thought timely for the Reith Lectures that year to have a scientific theme, and I was invited to give them by Mark Damazer, controller of BBC Radio 4. Mark's persuasive powers overcame my diffidence and my unease about the format; I'm grateful for the advice and encouragement he offered me at every stage. Three of my Royal Society colleagues, Peter Collins, Peter Cotgreave, and James Wilsdon, made helpful comments on the script. I'm grateful also to Sue Ellis, Kirsten Lass, Martin Redfern, and their colleagues at the BBC for editorial suggestions, and for ensuring that the production and presentation of the lectures in four different locations went off smoothly. And I'm indebted to Sue Lawley for the zest and professionalism with which she introduced each lecture and moderated the lively discussions afterward.

Andrew Franklin encouraged me to prepare this expanded text; I am grateful to him, and to Penny Daniel, Susanne Hillen, and Rukhsana Yasmin for editorial advice and support at Profile Books, and to Angela von der Lippe and Laura Romain at Norton and Carol Rose, copy editor, for help with the US edition.